AF743817

PLANT COMPETITION IN CROPPING SYSTEMS

Edited by **Daniel Dunea**

Plant Competition in Cropping Systems
http://dx.doi.org/10.5772/intechopen.71291
Edited by Daniel Dunea

Contributors

André Andres, Francisco Goulart, Fábio Schreiber, Ananda Scherner, Germani Concenço, Sai Sravan, M. Anowarul Islam, Dennis Ashilenje, Ömür Baysal, Ragıp Soner Silme, Eva Banowati, Satya Budi Nugraha, Daniel Dunea

© The Editor(s) and the Author(s) 2018
The rights of the editor(s) and the author(s) have been asserted in accordance with the Copyright, Designs and Patents Act 1988. All rights to the book as a whole are reserved by INTECHOPEN LIMITED. The book as a whole (compilation) cannot be reproduced, distributed or used for commercial or non-commercial purposes without INTECHOPEN LIMITED's written permission. Enquiries concerning the use of the book should be directed to INTECHOPEN LIMITED rights and permissions department (permissions@intechopen.com).
Violations are liable to prosecution under the governing Copyright Law.

(cc) BY

Individual chapters of this publication are distributed under the terms of the Creative Commons Attribution 3.0 Unported License which permits commercial use, distribution and reproduction of the individual chapters, provided the original author(s) and source publication are appropriately acknowledged. If so indicated, certain images may not be included under the Creative Commons license. In such cases users will need to obtain permission from the license holder to reproduce the material. More details and guidelines concerning content reuse and adaptation can be found at http://www.intechopen.com/copyright-policy.html.

Notice
Statements and opinions expressed in the chapters are these of the individual contributors and not necessarily those of the editors or publisher. No responsibility is accepted for the accuracy of information contained in the published chapters. The publisher assumes no responsibility for any damage or injury to persons or property arising out of the use of any materials, instructions, methods or ideas contained in the book.

First published in London, United Kingdom, 2018 by IntechOpen
IntechOpen is the global imprint of INTECHOPEN LIMITED, registered in England and Wales, registration number: 11086078, The Shard, 25th floor, 32 London Bridge Street
London, SE19SG – United Kingdom
Printed in Croatia

British Library Cataloguing-in-Publication Data
A catalogue record for this book is available from the British Library

Additional hard copies can be obtained from orders@intechopen.com

Plant Competition in Cropping Systems, Edited by Daniel Dunea
p. cm.
Print ISBN 978-1-78984-341-5
Online ISBN 978-1-78984-342-2

We are IntechOpen, the world's leading publisher of Open Access books

Built by scientists, for scientists

3,800+
Open access books available

116,000+
International authors and editors

120M+
Downloads

151
Countries delivered to

Our authors are among the
Top 1%
most cited scientists

12.2%
Contributors from top 500 universities

BOOK CITATION INDEX
CLARIVATE ANALYTICS
INDEXED

WEB OF SCIENCE™

Selection of our books indexed in the Book Citation Index
in Web of Science™ Core Collection (BKCI)

Interested in publishing with us?
Contact book.department@intechopen.com

Numbers displayed above are based on latest data collected.
For more information visit www.intechopen.com

For EU product safety concerns:
IN TECH d.o.o., Prolaz Marije Krucifikse Kozulić 3, 51000 Rijeka, Croatia,
info@intechopen.com or visit our website at intechopen.com.

Meet the editor

Daniel Dunea currently works as an Associate Professor and Head of Department in the Department of Environmental Engineering at Valahia University of Targoviste, Romania. He received his Ph.D. degree in Agronomy at the University of Agronomical Sciences and Veterinary Medicine Bucharest, Romania in 2007. From 2001 to 2002, he earned a Ph.D.20 Marie Curie Fellowship at Wageningen University and Research, Institute of Production Ecology & Resource Conservation in The Netherlands. He has published 5 books and 44 articles indexed in Clarivate Analytics Web of Science studying topics from environmental and agricultural sciences http://www.researchgate.net/profile/Daniel_Dunea/. He has reviewed more than 50 manuscripts for 17 journals with impact factor https://publons.com/author/1182105/daniel-dunea#profile. He has acted as scientific coordinator for several national and international research projects.

Contents

Preface

The idea of creating a book titled "Plant competition in cropping systems" came from the current context of conventional agricultural technologies that requires high quantities of chemical inputs and energy. Such practices are directly influencing climate change, water scarcity and quality, and environmental issues leading to health issues in humans and ecosystems. Effective management of natural resources can be encouraged by orienting the common agricultural practices towards the *functional biodiversity* concept in designing and implementing sustainable and eco-friendly cropping systems. This book provides basic principles and several case studies of polycrop utilization in various regions of the world as a method of functional biodiversity amplification through species associations that maximize the productivity per unit of land area, suppress the growth and development of weeds, and reduce the amount of harmful pests and insects.

This book provides knowledge on plant competition in multiple cropping systems, new insights regarding the ecological role of biodiversity for crop protection from pest management, soil fertility and plant health to plant resistance, and biodiversity indices to solve problems associated with legume persistence in cropping systems. It also discusses the benefits of nanoformulation of pesticides through target-oriented nanoparticles and their application against crop pests and diseases. Several case studies from Brazil and India (rice-based multiple cropping systems), and Indonesia (cassava cropping under the teak stands) are also presented. The inclusion of rice cultivars with greater competitive ability represents a promising tool for weed management in Brazil, since new cases of herbicide resistance are often reported and alternative control strategies are scarce. In India, inclusion of green manures/pulses/leguminous crops in nutrient exhausted rice-based cropping system saves the nitrogen fertilizer for the successive crops, increases the grain yields and profitability, and improves the soil structure. In Indonesia, the pattern of land utilization under the teak stands requires the selection of suitable plants according to the temporal dynamics, namely the season (dry or wet) and the plant's age.

I would like to thank the contributing authors for sharing their expertise on this promising topic and the IntechOpen editorial staff for providing specialized support in finishing this book.

I hope that this book will attract more enthusiasts in the study of the complex competition mechanisms occurring in polycrops, as well as the criteria that enable successful multiple cropping systems for intensive operations, and financial and environmental benefits. The mission is to combine all the intrinsic elements of profitable polycrops into a holistic approach that fully characterizes the *genotype – environment - management* interaction.

Daniel Dunea
Associate Professor Dr.
Head of Environmental Engineering Department
Valahia University of Targoviste, Romania

Section 1

Functional Biodiversity and Cropping Systems

Chapter 1

Introductory Chapter: Plant Competition in Multiple Cropping Systems beyond Conceptual Knowledge

Daniel Dunea

Additional information is available at the end of the chapter

http://dx.doi.org/10.5772/intechopen.81076

1. Introduction

In the coming years, farmers will face difficult challenges throughout the world in the context of climate change, water scarcity and environmental issues caused by conventional agricultural technologies. An effective management of natural resources can be encouraged by orienting the common agricultural practices towards the *functional biodiversity* concept in designing and implementing sustainable and eco-friendly cropping systems. It is well established that the enhancement of biodiversity facilitates and ameliorates the natural regulatory mechanisms of pests, insects and weeds [1].

The cropping practices that amplify the functional diversity and sustainability of an agroecosystem are as follows: diversification of farms, crop rotation, landscaping and polycrops (cultivation of various plants' associations). The polycrops are usually considered in low-input farming systems for weed control and optimization of the inputs, thus minimizing costs and the use of herbicides. Comparing with soled crop, multiple cropping systems provide faster propagation of canopy soil cover, improvement of absorbed photosynthetically active radiation (PAR), better competition with weeds and enhanced capture of available resources (light, water, nutrients). However, polycrops need to maintain at least the financial yield, crop quality and labour efficiency of the soled crops. Multiple cropping systems can be included in a scheme of relay crops. Therefore, the use of successive crops and intercrops provides an intensified use of land leading to better yields, optimized scheduling of harvests, assortment diversification and additional incomes.

Indeed, polycrops have shown a number of experimentally demonstrated advantages:

- Widen the productivity capacity of the arable land by maximizing the exploitation possibilities in time and space [2].

IntechOpen

© 2018 The Author(s). Licensee IntechOpen. This chapter is distributed under the terms of the Creative Commons Attribution License (http://creativecommons.org/licenses/by/3.0), which permits unrestricted use, distribution, and reproduction in any medium, provided the original work is properly cited.

- Provide suppression of weeds through niche preemption and interspecific competition for resources [3, 4].
- Support the complementarity of resource consumption from physiological, temporal and morphological point of view for the associated species [5].
- Ensure superior yields due to the efficient utilization of available resources, canopy space and the mutual interactions between heterogeneous canopy components [6].
- Repel insects and diminish pests' proliferation [7, 8].

The compatibility index of species that form the phytosociological associations relies on the degree of protocooperation within the interspecific competition, which influences the *net biological efficiency* of each species in the mixed canopy. The *net biological efficiency* (economic efficiency) is a fraction of the biological efficiency, and the ratio between them is known as the *harvest index*. The biological efficiency represents the total dry matter accumulation of the canopy, including the aerial and root biomass, starting from the emergence to the crop harvesting. Canopy architecture plays a strategic role in the association of species and must advantage the cash crop to capture PAR. Interspecific competition for light is an instantaneous process of resource capture, and the process efficiency is closely related to the light interception and light use characteristics of each species [6]. Height and leaf area distribution of both species are crucial for the canopy growth and development. Canopy structure and species growth are closely related because the structure results from the growth of individual plants within the canopy, thus affecting the rate of resource capture in the polycrop [5, 9]. Several factors related to the genetic traits of each species and to the technological factors influence the biological efficiency of species in the mixed canopy [10].

Two or more crops that are simultaneously grown on the same field must have adequate space to maximize their protocooperation and minimize the interspecific competition. Four basic elements need to be considered when designing polycrops [4]: spatial arrangement, density of plants, maturity period of the component species and the canopy architecture (**Figure 1**).

Most of the practical systems are variations of four basic spatial arrangements, such as row intercropping, strip intercropping, mixed intercropping and relay intercropping (**Figure 2**). A successful polycrop relies on (1) a detailed planning of the system, (2) sowing/planting in the optimal period for each associated species, (3) proper fertilization scheme, (4) integrated control of pests and insects and (5) efficient harvesting of each component species. The use of polycrops in the vegetable production systems (cover crops, intercrops, 'trap' crops, successive crops) brings benefits in terms of crop productivity and yield stability [11, 12]. Cereal-legume intercrops are among the most frequently used and most productive [13]. Although intercropping is less frequently used in high-input agricultural systems, mixtures of cereals (such as barley, wheat or oat) with forage legumes (such as white clover, red clover or alfalfa) are common in mechanized temperate farming systems providing the suppression of perennial weeds.

In the framework of polycrop science, the present book provides basic fundaments and several case studies of polycrop utilization in various regions of the world as a method of functional biodiversity amplification through species association that maximizes the productivity per unit of land area, suppresses the growth and development of weeds and reduces the populations of harmful pests and insects. Furthermore, the utilization of polycrops is a prospective instrument in evaluating arable land utilization options and in designing new

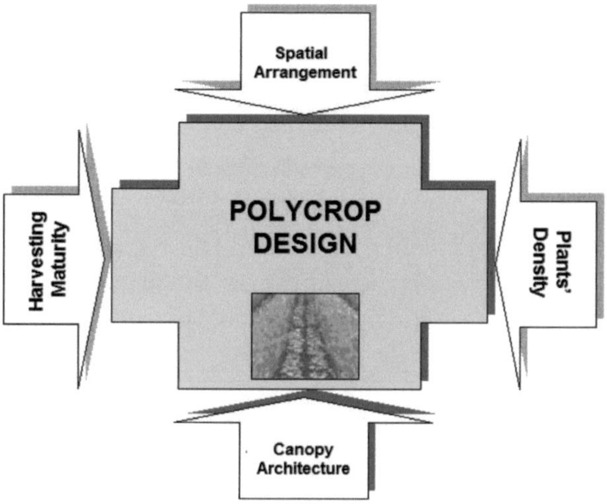

Figure 1. Criteria for designing polycrops.

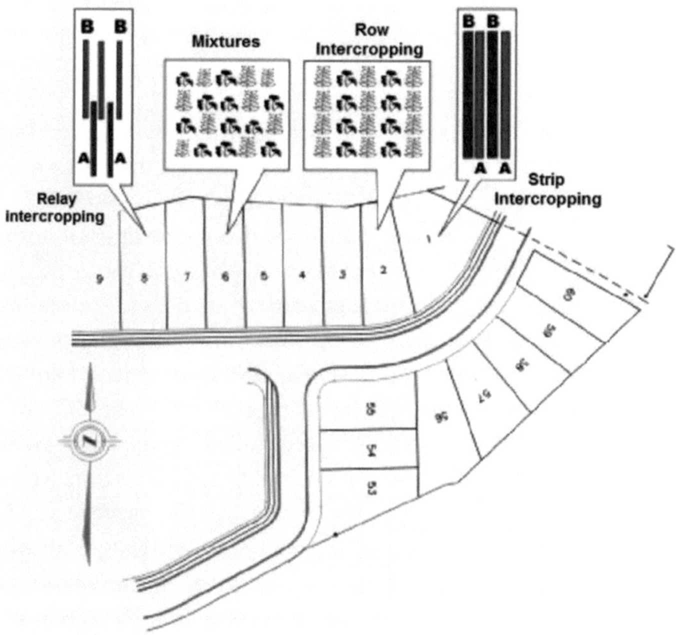

Figure 2. Spatial arrangements of polycrops.

cropping technologies, which provides sustainable cropping alternatives in the context of agroecosystem development at the ecoregional level.

The book is organized in six chapters that are divided in two sections which are as follows:

- Functional biodiversity and cropping systems: The first chapter introduces the reader to plant competition in multiple cropping systems. The second chapter provides insights regarding the ecological role of biodiversity for crop protection from pest management, soil fertility and plant health to plant resistance. It also discusses the benefits of nanoformulation of pesticides through target-oriented nanoparticles' (NPs) syntheses and their application against crop pests and diseases because they are cost-effective, nontoxic and environmentally friendly approaches. The third chapter analyses species traits and biodiversity indices to solve problems associated with legume persistence in cropping systems providing details regarding the competition for resources.

- Multiple cropping systems and plant competition: This section contains three chapters that present rice-based multiple cropping systems from Brazil and India and cassava cropping under the teak stands in Indonesia. The inclusion of rice cultivars with greater competitive ability represents a promising tool for weed management in Brazil, since new cases of herbicide resistance are often reported, and alternative control strategies are scarce. In India, the inclusion of green manures/pulses/leguminous crops in nutrient-exhaustive rice-based cropping system saves the nitrogen fertilizer for the successive crops, increases the grain yields and profitability and improves the soil structure. In Indonesia, the pattern of land utilization under the teak stands requires the selection of suitable plants according to the temporal dynamics, namely, the season (dry or rainy) and the plants' age.

2. Final considerations

Successful polycrops provide benefits for rural development by maximizing outputs (yields) and land equivalent ratio and minimizing inputs (fertilizers, herbicides and pesticides). Pest levels are often lowered in polycrops [14]. Farmers have generally regarded multiple cropping as a technique that reduces risks in crop production; if one member of an intercrop fails, the other survives and compensates in yield to some extent, allowing the farmer an acceptable harvest [15]. To gain acceptance, such agricultural practice must provide advantages over the other available options of the farmers. Obstacles in adoption of new strategies or practices of diversification are identified at sociological and financial level, rather than technological (a difficult step from conceptual to procedural knowledge). However, further research is still needed to assess the mechanisms of competition between species, to establish suitable companion species and to conceive intensive sequences of operations and adapted mechanization. Most of the polycrops are more suitable for extensive practices on small farms, but the move towards organic farming can compensate the production losses through higher prices of the agricultural ecological products. A keen extension strategy is necessary to familiarize the farmers with successful multiple cropping systems. To introduce such systems at farm level, first steps would imply trial fields located in the rural area to show the potential benefits of multiple cropping to the farmers.

Author details

Daniel Dunea

Address all correspondence to: dan.dunea@valahia.ro

Faculty of Environmental Engineering and Food Science, Department of Environmental Engineering, Valahia University of Targoviste, Targoviste, Romania

References

[1] Sullivan P. Intercropping Principles and Practices. 2001. http://www.attra.org/attra-pub/PDF/intercrop.pdf

[2] Gliessman SR. Agroecology: The Ecology of Sustainable Food Systems. Boca Raton: CRC Press; 2006

[3] Liebman M. Ecological Suppression of Weeds in Intercropping Systems. Boca Raton, FL: CRC Press; 1988

[4] Teasdale JR. Cover plants, smother plants, and weed management. In: Integrated Weed and Soil Management. Chelsea, MI: Ann Arbor Press; 1998

[5] Dunea D. Bioconversion Efficiency in Grass-Legume Forage Systems. Vol. 1. Saarbrücken: LAP Lambert Academic Publishing; 2015. p. 256

[6] Kropff MJ, van Laar HH. Modelling Crop-Weed Interactions. Wallingford: CAB International; 1993

[7] Altieri MA, Leibman M. Insect, weed, and plant disease management in multiple cropping systems. In: Francis CA, editor. Multiple Cropping Systems. New York: Macmillan Company; 1994. p. 383

[8] Altieri MA. The ecological role of biodiversity in agroecosystems. Agriculture, Ecosystems and Environment. 1999;**74**:19-31

[9] Keating BA, Carberry PS. Resource capture and use in intercropping: Solar radiation. Field Crops Research. 1993;**34**:273-301

[10] Aldrich RJ. Weed-Crop Ecology. Principles in Weed Management. North Scituate, MA (USA): Breton Publishers; 1984

[11] Baumann DT. Competitive suppression of weeds in a leek-celery intercropping system [PhD thesis]. Wageningen University; 2001

[12] Coaker TH. Cultural methods: The crop. In: Brun AJ, Coaker TH, Jepson PC, editors. Integrated Pest Management. London: Academic Press; 1987. pp. 69-88

[13] Ofori F, Stern WR. Cereal-legume intercropping systems. Advances in Agronomy. 1987; **41**:41-90

[14] Fukai S, Trenbath BR. Processes determining intercrop productivity and yields of component crops. Field Crops Research. 1993;**34**:247-271

[15] Bowen JF, Kratky BA. Successful Multiple Cropping Requires Superior Management Skills. Agribusiness Worldwide. November/December. 1986. pp. 22-30

Chapter 2

The Ecological Role of Biodiversity for Crop Protection

Ömür Baysal and Ragıp Soner Silme

Additional information is available at the end of the chapter

http://dx.doi.org/10.5772/intechopen.78228

Abstract

Agricultural system is a complex community sheltering different ecological units. The units of this complex structure are in balance with each other showing fluctuations to ensure effective regulations from time to time depending on the abundance of both undesirable and beneficial organisms. This balance is a major case for biological activity playing an important role to maintain biological diversity. Once this natural balance is impaired due to abiotic and biotic factors occurring in biosystems, the economic and environmental problems appear becoming significant for the economical dimension in agriculture. The most important components showing deficiencies in systemically agro-ecostructure problems result from soil fertility, pest and disease management. Large interactions, which are concomitantly persisting with biological processes, are on plant and animal biodiversity, which have been affected by miss-treatments in crop protection and plant nutrition. Hence, food-web and biodiversity are indirectly seriously damaged in nature, such as recycling of nutrients and changes of microclimate. In this chapter, we have discussed the major effects of crop protection on biodiversity in detail regarding the persistence of biodiversity that needs to be mediated, considering the preserving of ecological properties and sustainable maintenance of biological integrity in agroecosystems.

Keywords: agroecology, antagonists, biodiversity, biological control, target-oriented nanotechnological approaches, environment-friendly approach, sustainability

1. Introduction

As a main value of the nature, biodiversity refers to all living species existing and interacting within an ecosystem and within each other such as microorganisms, plants, animals etc. [1]. It has also a major role as a source of agricultural production and cultivation of domestic crops. Breeding and hybridization techniques are efficient ways to increase their yield and quality that seems valuable genetic resources for crop improvement, which serves indirectly and

IntechOpen

© 2018 The Author(s). Licensee IntechOpen. This chapter is distributed under the terms of the Creative Commons Attribution License (http://creativecommons.org/licenses/by/3.0), which permits unrestricted use, distribution, and reproduction in any medium, provided the original work is properly cited.

directly for many ecological cases. In agricultural systems, this unique property of ecosystem provides food sources and organic fuels besides web of nutrients, regulation of microclimatic conditions, ongoing required hydrological processes, removal of undesirable residues of macro/micro-organisms, and hazardous chemicals. The recycling and renewing processes largely occur biologically being directly dependent on the existence of biological diversity [2]. Once this natural process is impaired, the losses in economic and environmental fields will be seriously significant. The lack of functional components and properties of soil fertility and pest regulation reduces the quality of life due to contaminated soil, water, and food quality by pesticide and/or nitrate accumulations.

For creating an artificial ecosystem through nutrient recycling supplied by only chemical fertilization and control of pest and pathogens by chemical pesticides, results in constant but infertile and not sustainable ecosystems, which are used for agricultural purposes created by human intervention. In fact, it is an inevitable end for the functional regulation of nature by impairment of biodiversity, which will extinct the flows of energy and the nutrients will progressively diminish because of the intensive crop cultivation [3].

In our century, seedling preparation and mechanized planting have replaced the conventional methods. Genetic manipulations have been used in breeding and selection of varieties. Modern agricultural systems bring high incomes depending on external inputs. Many differences of opinions are present concerning the protection of non-renewable resources, the loss of biodiversity, the loss of land by soil erosion and lack of biological property by chemical fertilizers and pesticides that have negative effects on human and animal health, food quality and safety, and environmental pollution [4].

Nowadays, increasing in pollution of environmental conditions enforces us to develop agroecological ecofriendly approaches considering the conservation of biodiversity, soil, water and other resources that is an inevitable requirement for sustainable preservation of environmental structure in the world. Therefore, enhancing of functional biodiversity is a key strategy for living ecosystems including beneficial antagonists and soil microflora dynamism in crop protection and soil fertility [5, 6].

2. Biodiversity in agroecosystems

Modern agriculture enforces the use of all components of nature available to human beings that determine the simplification of nature's diversity considering a diminished number of cultivated plants and domesticated animals. The literature and other knowledge sources indicate that only a few species of grain, vegetable, and fruit crop species are intensively cultivated [7] besides the huge diversity of plant species found in tropical rain forest containing nearly 100 species of trees (**Figure 1**) [4, 8]. Genetically, modern agriculture is under the pressure of major crops limiting varieties in cultivated areas [9] that creates genetic uniformity and determines day-by-day losses in biodiversity.

The conventional crop cultivation system consists of different varieties of domesticated crop species and their wild relatives showing full or partial resistance to diseases that allows farmers to produce crops in different soil types and microclimates [10].

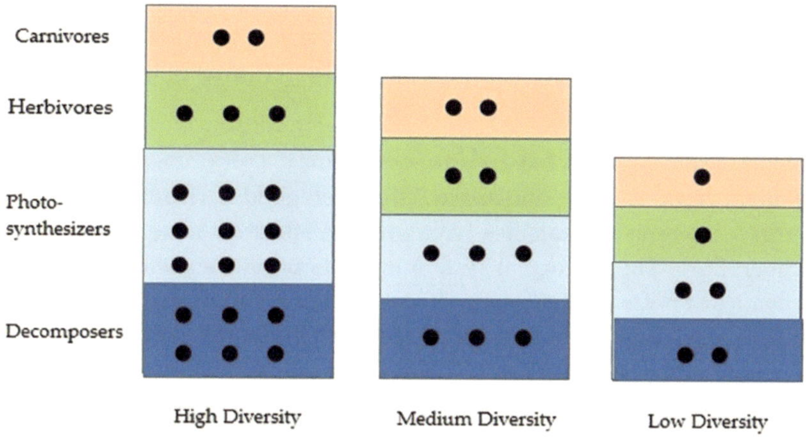

Figure 1. Simple diagram showing the difference between high diversity and low diversity.

In general, the degree of biodiversity depends on four main characteristics of the agroecosystem [4, 11]:

i. The diversity of vegetation within and around the agroecosystem.

ii. The stable maintenance and permanence of various crops.

iii. The intensity of crop cultivation activities and pest control management.

iv. The divergent parts of the agroecosystem from natural vegetation.

The biodiversity part of an agro-ecosystem can be clustered according to their role in cropping systems. It contributes to the productivity through pollination, biological control, degradation besides components such as weeds, insect pests and microbial pathogens. Ecological key is to identify the type of biodiversity that is desirable to maintain and/or enhance the best practices that will encourage the formation of biodiversity components [4]. Many agricultural practices have the potential to enhance functional biodiversity, besides the artificial manipulation that is negatively affecting the ones mentioned above. The main idea is to select the best management practices to enhance or regenerate this kind of biodiversity such as nutrient cycling, water and soil conservation, biological pest management, etc.

3. Biodiversity and pest management

Because of biodiversity reduction, unconscious pesticide applications and mistreatments of soils are shown as main reasons. One problem in agroecosystems is increasingly correlated to monocultures and decreasing of diversity [12]. Plant varieties that are modified to meet the special requirements of consumers are under attack of heavy pests' damage [13]. The characteristic properties depending on trait locus of natural communities are lost by exogenous modifications. The literature on biodiversity suggests that the design of vegetation management strategies must include knowledge on crop arrangement in time and space, the composition and abundance of non-crop vegetation within and around fields, and the soil type including its environment and intensity of management. Extension of the cropping period or planning cropping frequency may allow naturally-occurring biological control agents to sustain higher population levels on alternate hosts and to persist in the agricultural environment throughout the whole season [4, 13, 14].

Low pest potentials may be expected in agro-ecosystems if a production area exhibit high crop diversity by mixing crops in time and space. Moreover, good agricultural practices including integrated crop management strategies have positive effect on remediation of characteristic property of microflora. The ecological system in fields provides shelter and alternative food for natural enemies of pests. Pests may proliferate in these environments depending on population dynamism of natural enemies/or presence of alternate hosts in the area. Orchards are, in some extent, permanent ecosystems, and they are more stable than annual crops. They have greater structural diversity, possibilities for the establishment of biological control agents by floral diversity conditions. Increase of crop densities or cultivation of tolerable specific weed species is a bio-remediation tool for biodiversity combined with the use of variety mixtures or crops. These are few prominent properties that are necessary in the planning of a crop management strategy in agroecosystems.

4. Biodiversity, soil fertility and plant health

To understand the main factors of plant biodiversity, climate and geographic properties should be considered at the micro-fauna level. The relationship between plant biodiversity and productivity can also be influenced by other abiotic and biotic factors (**Figure 2**). Soil

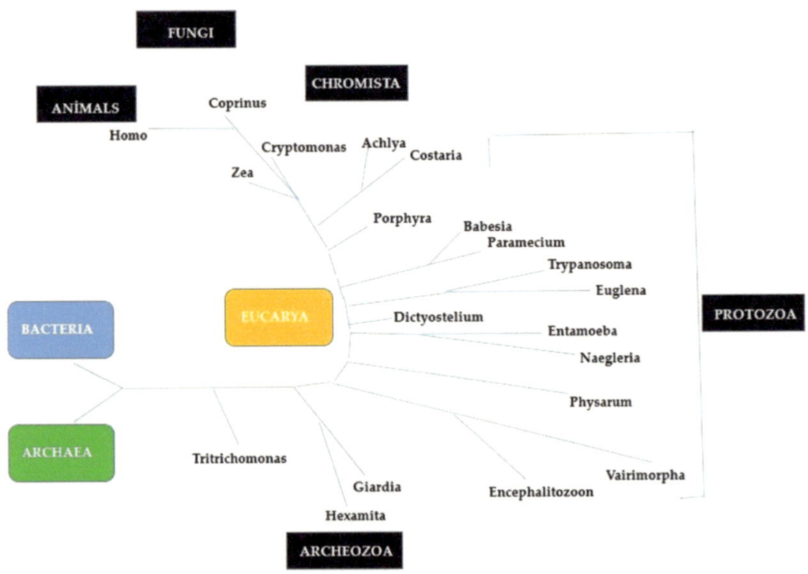

Figure 2. Simple diagram showing biotic factors through the evolutionary approach.

biodiversity reductions occur from negative issues due to recycling of nutrients and improper balance between organic matter, soil organisms and plant diversity. These are necessary components of a productive and ecologically-balanced soil environment [4, 15–17]. Soil biomass consists of beneficial and harmful microbes (fungi, bacteria and actinomycetes) and animals such as nematodes and different insects. One gram of soil contains nearly a thousand fungal hyphae and a million bacterial colonies [3]. Soil organisms provide a number of vital functions [18]: degradation of litter and cycling nutrients converting atmospheric nitrogen into organic forms, and reconverting organic nitrogen besides suppressing soil-borne pathogens through antagonism, synthesize of enzymes, vitamins, hormones, vital chelators altering soil structure through population living in mutualistic, commensalistic, competitive, and pathogenic forms.

The microbial activity of soil directly and/or indirectly affects the nutrient availability and plant nutrition. Decomposition of organic matter by microbial activity is used in cell building and maintenance processes of plants that are sources of available nutrients for plants. Furthermore, because of the microbial competition ongoing at different fractions of the soil organic matter, nutrients in biomass secreted compounds and dead cell of microorganisms are attacked by other competitive microbial communities. The effect of microbial activity has a positive effect on the available form of nutrients and elements that increase plant resistance to pathogens [19].

4.1. Positive reflection of dynamic biota on soil fertility and plant resistance

Many studies show that biologically suppressive activity can be regulated by the physical and chemical characteristic properties of soils [20–22]. Disease-conducive soils are described as a living biomass that is insufficient showing no suppressive effect on pathogens. In biologically balanced microflora, the disease-suppressive effects of microbes are successfully manipulated in order to suppress pathogens and thereby, they reduce disease losses [23]. The mechanisms are in most cases not well known, the manipulation of soil biological activity and enhancing biodiversity appear to be a method by which pathogen invasion on plant can be reduced. Studies have revealed novel antagonistic relationships between soil organisms and soil-borne pathogens [24, 25] and identified methods by which the soil environment can be manipulated to suppress pathogen activity [22, 26]. Pathogen-inhibitory components secreted and released may act against pathogens, which have fungistatic or fungicidal properties [27].

Some studies have reported interactions based on untested and often postulated and/or unstated assumptions; we simply do not have enough information on microbial dynamism and ongoing struggle to survive in soil microflora to successfully use disease-suppressive microbes in a wide variety of cropping environments. However, many researchers have suggested a direct relationship between soil biodiversity and disease suppression depending on dynamic population that will increase suppressive effect of the soil that regulates this dynamism. Such information is critical for the understanding of these relationships and the testing of whole assumptions. Because of the demonstrable dynamism of microbial population on the food supply, soil-borne plant pathogens provide useful models for evaluation of the impact of soil biodiversity on agroecosystems [22].

As a new concept, biological control of plant pathogens can be realized by using of inoculations and introduction of effective microbial species that are protecting our crops from plant pathogens' attacking and establishment of safety microflora based on introduced organisms. Inoculation of seeds with biocontrol agents and/or dipping of roots into solution of antagonistic microbes (*Rhizobia, Mycorrhizae,* and *Trichoderma*) have a direct protective effect to enhance plant performance and resistance to pathogens [28]. When pathogens are not inhibited by naturally present antagonists, it is possible to enhance biocontrol by adding more effective ones selected by previous studies and data relying on scientific evaluations. For instance, *Agrobacterium tumefaciens* var. *radiobacter* strain 84 and *Peniophora gigantea* have been successfully introduced and used against crown gall (*Agrobacterium tumefaciens*) in fruit trees. Many other tested microorganisms inhibiting pathogens have positive effect on plant health and induction of resistance when introduced into the soil or plant rhizosphere e.g., *Trichoderma* spp., *Pseudomonas* spp., *Bacillus* spp., *Alcaligenes* spp., *Agrobacterium tumefaciens* and others [4, 25, 29].

The biocontrol aims to introduce antagonistic microorganisms in soil, without considering the nutrient content of soil, to diminish pathogen population thereby adversely affecting infection process. A number of fungal and bacterial parasites can be used to control of most destructive soil-borne nematodes (*Meloidogyne* spp). There are many ways in which an antagonist microorganism can show rapid colonization in advance of the pathogens. Competition between biocontrol agent and pathogen may lead to niche exclusion, secretion of secondary metabolites and/or antibiotics may create an unsuitable medium resulting in cell-wall degradations of the pathogen. In addition, some microorganisms positively induce growth of plants, so that even if disease is present, its symptoms are partly masked. Moreover, ectomycorrhizae promotes phosphorous uptake in plants, forming a physical layer or a chemical barrier to pathogen invasion, thereby preventing pathogens from affecting the root surface of a plant [4, 30]. The literature on soil management recommends the enhancing of existing microbial antagonists, use of organic amendments reported as initiators of disease control processes to provide appropriate conditions for secreting of metabolites with digestive compounds by soil microorganisms [31]. Organic additions have an active role on microbial activity and supply advantages to antagonistic individuals in controlling of pathogens [32].

5. Target-oriented nanotechnological approaches and preservation of biodiversity

In the past decades, chemical pesticides have been widely used for plant protection. Nevertheless, hazardous chemicals are not only affecting the target pest but also other natural enemies modifying the biological balance. The negative effects of chemicals and residues have become also a public concern since they cause health disorders and environmental pollution. Therefore, nano-formulation of these chemicals has received much attention to diminish of these side effects. Target oriented nanoparticles (NPs) syntheses and their application against crop pests and diseases have been suggested since they are cost-effective, non-toxic and environmentally friendly biological approaches [33]. Converting of metallic compounds into nanoparticle forms increases its effect on target pathogen and pest. Hence, we are able to reduce the side effect of hazardous components and source of these chemical components are used for pest and pathogen control. Moreover, nanotechnology has been used for detection of plant pathogens using biosensor-based synthesized products [34]. Different nano-formulations of these molecules have been proposed since they provide efficient identification and effective management considering the biosafety and preservation of biodiversity.

It is essential to understand the biochemical and molecular mechanisms of nanoparticle synthesis. They have been suggested due to its long lasting biological activities compared to conventional pesticides. Besides their multifaceted property enhancing the volume ratio, it reduces the amount of pesticide to be used and provides better contact on target surface. However, recent studies have shown that there are some negative effects on biodiversity [35].

Advanced agronomical methods enforce agricultural production through the use of effective fertilizers and pesticides based on nanotechnology. However, their negative effects in the ecosystem have indirectly influenced the biological diversity and contaminate groundwater and soil [36].

Green nanotechnology has two objectives: creating nanomaterials and items without hurting the Earth or human wellbeing, and delivering nano-items giving answers for ecological issues. It utilizes existing standards of green science and green designing [37] and make nanomaterials and nano-items without poisonous fixings, at low temperatures as less vital and inexhaustible sources by considering lifecycle thinking in all outline and designing stages. Administrative bodies, for example, the United States Ecological Assurance Organization and the Sustenance and Medication Organization in the U.S. and the Wellbeing and Insurance Directorate of the European Commission have begun the managing of potential dangers generated by nanoparticles. Constrained nanotechnology and control are necessary for potential human and ecological wellbeing and security issues related to nanotechnology. It has been contended that the improvement of far reaching control of nanotechnology will be indispensable even we are able to determine of their potential dangers related to the examination and business utilization besides potential advantages [38].

Nanotechnology has diverse applications in precision agriculture. However, toxicity can be a major problem of nanoparticles due to their unique properties. Effects of the unique characteristics of nanoparticles are not well understood; hence more studies on toxicity are required for commercial food crop applications [39]. However, applications of nanoparticles are not always detrimental to plants and they have also positive effects [40–42].

Carbon nanomaterials such fullerenes, carbon nanoparticles, fullerol, and single-walled carbon nanotubes/multiwall carbon nanotubes have been utilized as a part of agribusiness demonstrating positive and unfavorable impacts. Lethality of carbon nano-materials was observed to be to a great extent reliant on their fixations, development/presentation conditions, and plant species. Kerfahi et al. [43] examined the impacts of local and functionalized multiwall carbon nanotubes (0–5000 mg/kg) on soil microbes. They revealed that following 2 weeks, the dirt bacterial group was significantly influenced by the multiwall carbon nanotubes. Following 2 months, there was no impact on the bacterial assorted variety with either kind of nanotubes. They ascribed this early impact to the acidic behavior of multiwall carbon nanotubes that caused a diminishing in soil pH at higher introduction fixations and hence changed the soil bacterial groups [43].

In another study, Boonyanitipong et al. [44] considered phytotoxicity of zinc oxide and titanium dioxide nanoparticles on rice (*Oryza sativa* L.) roots. The following three parameters were investigated: seed germination rate, root length and number of roots. The outcomes demonstrated that there was no decrease in the percent seed germination from zinc and titanium dioxide nanoparticles. However, zinc oxide nanoparticles demonstrated hindering growth of rice roots at the early seedling stage. This examination demonstrated that immediate introduction without pre-testing of particular kinds of nanoparticles could cause critical phytotoxicity and accentuated the need for biologically controlled transfer of wastes containing nanoparticles and further use in horticultural and ecological setups [44]. Chai et al. [45] considered the impact of metal oxide nanoparticles (ZnO, SiO_2, TiO_2 and CeO_2) on useful microbes and metabolic profiles in horticultural soil. ZnO and CeO_2 nanoparticles led to the obstruction of thermogenic digestion, diminished the quantities of *Azotobacter*, P-solubilizing and K-solubilizing microbes in soil and restrained the enzymatic activities [45].

These studies showed that nanotechnological approaches should be carefully used considering their adverse effect on biodiversity and population dynamism of micro/macro organism besides its positive sides.

6. Outlook and future aspects

In brief, the beneficial opportunities of microorganisms have been mentioned in literature and published reports. Nevertheless, artificially mimicking of their activities by present technology is impossible when estimating turnover time of biomass, which is 1000–10,000 times less than that obtained in optimal *in vitro* conditions [46]. The data suggests only active short periods and dormant state in soil for microorganisms, which are able to survive in harsh conditions [47]. Technical limitations have made difficult to follow and understand the mysteries of microflora, in all cases with any degree of confidence. The recent application of molecular technologies will revolutionize this scientific area and may permit us to gain a more complete understanding of soil biodiversity [48]. With this information, the use of cultural practices to manipulate microbial activity and diversity may become more practical and effective for the management of soil-borne diseases [22]. Further, there is a need for a better understanding of the capacity of soil-borne pathogens to generate new biotypes depending on phenotypic variation (**Figure 3**) in response to selection pressures, to improve effectively the pest control. New molecular technologies such as „metagenomics "provide great opportunities for precise measurement of both soil biodiversity and pathogen variability. These tools can be used to directly test hypotheses concerning the interactions between soil organisms and plant pathogens.

Efficient and effective protocols for extraction, characterization and quantification of soil DNA and RNA, besides new disease-resistant cultivars, including employing new resistance strategies have been developed using modern biotechnology [49]. Particularly, in assessing soil biodiversity which has potential to suppress soil-borne pathogens, e.g., "metagenomics" can be used (**Figure 4**), that these analyses will be beneficial for the comprehensive understanding of the traits of microbes, which are normally very difficult to measure of their biogeochemical property or potential effect of non-cultivated ones at micro-scale using conventional microbiological methods. To our knowledge, such advanced tools have not yet been used to directly compare microbial metagenomes across soils representing a range of different biomes.

http://dx.doi.org/10.5772/intechopen.78228

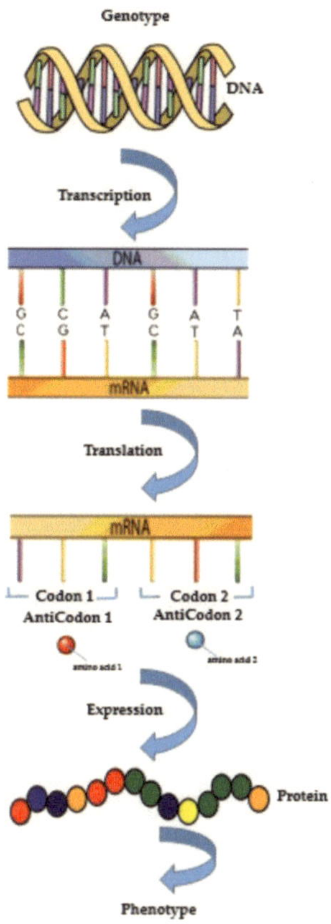

Figure 3. Simple diagram shows that how genetic diversity affects biological diversity based on phenotype formation depending on genotype.

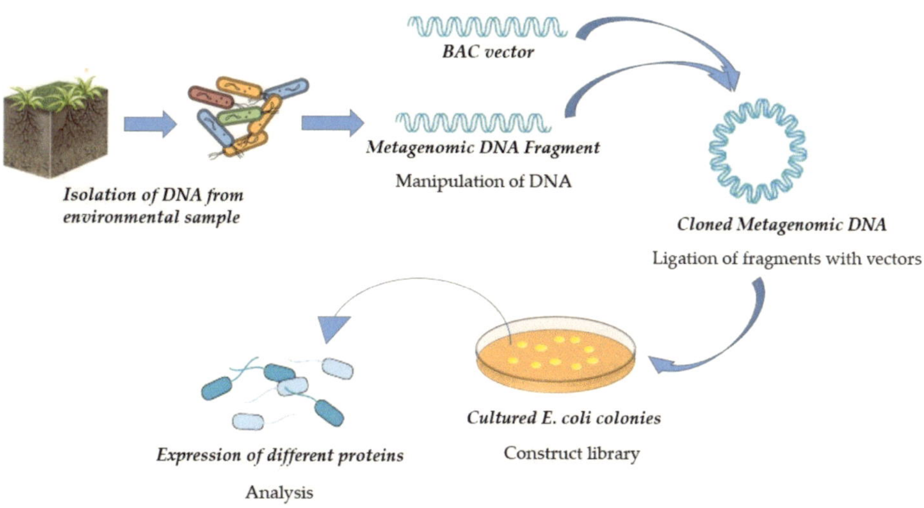

Figure 4. Simple diagram showing the workflow of metagenomics.

"Metagenomics" is a systematically investigation method for classifying and manipulating the entire genetic material isolated from environmental samples. This contain a multi-step process that relies on the efficiency of four main steps consisting of the isolation of genetic material, manipulation of the genetic material, library construction, and the analysis of genetic material in the metagenomics library. Information from metagenomics libraries has the ability to enrich the knowledge and applications of many aspects of environmental sustainability and remediation of soil property. This information can be applied to create a healthy and dynamic microbial population that lives in balance with the environment. Metagenomics is an efficient tool and an exciting field of molecular biology that is likely to grow into a standard technique for understanding the biological diversity at advanced level.

7. Conclusion

We propose several methods with a measurable aspect that can provide benefits for soil biodiversity and may provide information for maintaining biodiversity. This information has also positive effect on plant pathology bringing a new improvement of by using molecular tools such as PCR and microarrays to quantify microbes and monitor gene expression and metagenomics. We believe that future data will provide more information than the previously available ones. Novel agro-ecological approaches will aim at breaking the negative effects of miss-applications related to integration of new plant protection techniques that enhance complex interactions and synergisms and optimize ecosystem functions and processes, such as biotic spontaneously regulation of harmful microorganisms, nutrient recycling, and biomass production and accumulation.

In short, considerable efforts and new technologies are needed to access not only DNA pools but also an entire metagenome for unbiased microbial ecology studies for both understanding and decipher the ecosystem mechanisms and for learning the most effective and eco-friendly control measurements to deal with pest and pathogens.

Conflict of interest

The authors have no conflict of interest.

Author details

Ömür Baysal[1]* and Ragıp Soner Silme[2]

*Address all correspondence to: omurbaysal@mu.edu.tr

1 Department of Molecular Biology and Genetics, Faculty of Science, Muğla Sıtkı Koçman University, Muğla, Turkey

2 Istanbul University, Istanbul, Turkey

References

[1] Vandermeer J, Perfecto I. Breakfast of Biodiversity: The Truth about Rainforest Destruction. Oakland: Food First Books; 1995. 185 p

[2] Altieri MA. Biodiversity and Pest Management in Agroecosystems. New York: Haworth Press; 1994. 185 p

[3] Swift MJ, Anderson JM. Biodiversity and ecosystem function in agroecosystems. In: Schultze E, Mooney HA, editors. Biodiversity and Ecosystem Function. New York, Spinger; 1993. pp. 57-83

[4] Altieri MA. The ecological role of biodiversity in agroecosystems. Agriculture, Ecosystems and Environment. 1999;**74**:19-31

[5] Kropff MJ, van Laar HH: Modelling Crop-Weed Interactions. Wallingford, U.K.: CAB International (in association with the International Rice Research Institute); 1993. 274 p

[6] Dunea D, Dincă N. Improving land utilization using intensive grass-clover mixtures in forage production systems. Romanian Agricultural Research. 2014;**31**:147-158

[7] Fowler C, Mooney, P. Shattering: Food, Politics and the Loss of Genetic Diversity. Tucson, AZ: University of Arizona Press; 1990. 278 p

[8] Perry, DA. Forest Ecosystems. Baltimore, MD: Johns Hopkins University Press; 1994. 649 p

[9] National Academy of Sciences. Genetic Vulnerability of Major Crops. Washington, DC: NAS; 1972. 307 p

[10] Brush SB. The natural and human environment in the Central Andes. Mountain Research and Development. 1982;**2**:14-38

[11] Southwood RE, Way MJ. Ecological background to pest management. In: Rabb RC, Guthrie FE, editors. Concepts of Pest Management. Raleigh, NC: North Carolina State University; 1970. pp. 6-29

[12] Altieri MA, Letourneau DK. Vegetation management and biological control in agroecosystems. Crop Protection. 1982;**1**:405-430

[13] Brasier CM. Episodic selection as a force in fungal microevolution, with special reference to clonal speciation and hybrid introgression. Canadian Journal of Botany. 1995; **73**:1213-1221

[14] Hsiang T, Mahuku GS. Genetic variation within and between southern Ontario populations of *Sclerotinia homoeocarpa*. Plant Pathology. 1999;**48**:83-94

[15] Hendrix, PF, Crossley, DA, Blair, JM, Coleman, DC. Soil biota as components of sustainable agroecosystems. In: Edwards, CA, Lal, R, Madden, P, Miller, RH, House, G, editors. Sustainable Agricultural Systems. IA: Soil and Water Conservation Society; 1990. p. 637-654

[16] Parkinson D. Linkages between resource availability, microorganisms and soil invertebrates. Agriculture, Ecosystems and Environment. 1988;**24**:21-32

[17] Durall DM, Parkinson D. Initial fungal community development on decomposing timothy (*Phleum pratense*) litter from a reclaimed coal-mine spoil in Alberta, Canada. Mycological Research. 1991;**95**:14-18

[18] Paoletti MG, Foissner W, Coleman D, editors. Soil Biota, Nutrient Cycling, and Farming Systems. Boca Raton: Lewis Publishers, 1994. 314 p

[19] Baysal Ö, Tör M. Smart biologics for crop protection in agricultural systems. Turkish Journal of Agriculture and Forestry. 2014;**38**:723-731. DOI: 10.3906/tar-1309-26

[20] Liu LX, Hsiang T, Carey K, Eggens JL. Microbial populations and suppression of dollar spot disease in creeping bentgrass with inorganic and organic amendments. Plant Disease. 1995;**79**:114-147

[21] Gracia-Garza JA, Bailey B, Paulitz TC, Lumsden RD, Reeleder RD, Roberts D. Enzymatic activity of *Trichoderma hamatum* grown on sclerotia of *Sclerotinia sclerotiorum* damaged by fungus gnats. Biocontrol Science and Technology. 1997;**7**:401-413

[22] Reeleder RD. Fungal plant pathogens and soil biodiversity. Canadian Journal of Soil Science. 2003;**83**(Special Issue):331-336. DOI: doi.org/10.4141/S01-068

[23] Chen WD, Hoitink HA, Schmitthenner AF, Tuovinen OH. The role of microbial activity in suppression of damping-off caused by *Pythium ultimum*. Phytopathology. 1988; **78**:314-322

[24] Baysal Ö, Çalışkan M, Yeşilova Ö. An inhibitory effect of a new *Bacillus subtilis* strain (EU07) against *Fusarium oxysporum* f. sp. *radicis lycopersici*. Physiological and Molecular Plant Pathology. 2008;**73**:25-32

[25] Baysal Ö, Lai D, Xu HH, Siragusa M, Çalışkan M, Carimi F, Teixeria da Silva JA, Tör M. A proteomic approach provides new insights into the control of soil-borne plant pathogens by *Bacillus* species. PLoS One. 2013;**8**(1):e53182:1-e5318212. DOI: 10.1371/journal.pone.0053182-2013

[26] Lazarovits G, Conn KL, Potter J. Reduction of potato scab, *verticillium* wilt, and nematodes by soymeal and meat and bone meal in two Ontario potato fields. Canadian Journal of Plant Pathology. 1999;**21**:345-353

[27] Lazarovits G. Management of soil-borne plant pathogens with organic soil amendments: A disease control strategy salvaged from the past. Canadian Journal of Plant Pathology. 2001;**23**:1-7

[28] Miller RH. Soil microbiological inputs for sustainable agriculture. In: Edwards CA, Lal R, Madden P, Miller RH, House G, editors. Sustainable Agricultural Systems. IA: Soil and Water Conservation Society; 1990. pp. 614-623

[29] Campbell R. Biological Control of Microbial Plant Pathogens. Cambridge: Cambridge University Press; 1989. 199 p

[30] Tjamos EC, Papavizas GC, Cook RJ. Biological Control of Plant Diseases: Progress and Challenges for the Future. NATO ASI Series, New York: Plenum Press; 1992. 462 p

[31] Palti J. Cultural Practices and Infectious Crop Diseases. New York: Springer; 1981. 243 p

[32] Fry WE. Principles of Plant Disease Management. New York: Academic Press; 1982. 378 p

[33] Thakkar KN, Mhatre SS, Parikh RY. Biological synthesis of metallic nanoparticles. Nanomedicine. 2010;**6**(2):257-262. DOI: 10.1016/j.nano.2009.07.002

[34] Fang Y, Ramasamy RP. Current and prospective methods for plant disease detection (Rains GC, editors). Biosensors. 2015;**5**(3):537-561. DOI: 10.3390/bios5030537

[35] Simonin M, Richaume A. Impact of engineered nanoparticles on the activity, abundance, and diversity of soil microbial communities: A review. Environmental Science and Pollution Research. 2015;**22**(18):13710-13723. DOI: doi.org/10.1007/s11356-015-4171-x

[36] Liu HH, Cohen Y. Multimedia environmental distribution of engineered nanomaterials. Environmental Science & Technology. 2014;**48**:3281-3292. DOI: 10.1021/es405132z

[37] Bonilla CAM, Kouznetsov VV. In: Larramendy M, editor. "Green" Quantum Dots: Basics, Green Synthesis, and Nanotechnological Applications, Green Nanotechnology - Overview and Further Prospects. London: InTech; 2016. DOI: 10.5772/62327. Available from: https://www.intechopen.com/books/green-nanotechnology-overview-and-further-prospects/-green-quantum-dots-basics-green-synthesis-and-nanotechnological-applications

[38] Shatkin JA. Nanotechnology: Health and Environmental Risks. 1st ed. Boca Raton, FL: CRC Press; 2008. 192 p

[39] Mattiello A, Filippi A, Pošćić F, Musetti R, Salvatici MC, Giordano C, Vischi M, Bertolini A, Marchiol L. Evidence of phytotoxicity and genotoxicity in *Hordeum vulgare* L. exposed to CeO_2 and TiO_2 nanoparticles. Frontiers in Plant Science. 2015;**6**:1043. DOI: 10.3389/fpls.2015.01043

[40] Ghrair AM, Ingwersen J, Streck T. Immobilization of heavy metals in soils amended by nanoparticulate zeolitic tuff: Sorption-desorption of cadmium. Journal of Plant Nutrition and Soil Science. 2010;**173**:852-860. DOI: 10.1002/jpln.200900053

[41] Khot LR, Sankaran S, Maja JM, Ehsani R, Schuster EW. Applications of nanomaterials in agricultural production and crop protection: A review. Crop Protection. 2012;**35**:64-70. DOI: 10.1016/j.cropro.2012.01.007

[42] El-Temsah YS, Oughton DH, Joner EJ. Effects of nano-sized zero-valent iron on DDT degradation and residual toxicity in soil: A column experiment. Plant and Soil. 2014;**368**:189-200. DOI: 10.1016/j.chemosphere.2013.02.039

[43] Kerfahi D, Tripathi BM, Singh D, Kim H, Lee S, Lee J, Adams JM. Effects of functionalized and raw multi-walled carbon nanotubes on soil bacterial community composition. PLoS One. 2015;**10**:e0123042. DOI: 10.1371/journal.pone.0123042

[44] Boonyanitipong B, Kositsup B, Kumar P, Baruah S, Dutta J. Toxicity of ZnO and TiO_2 nanoparticles on germinating rice seed *Oryza sativa* L. International Journal of Bioscience, Biochemistry and Bioinformatics. 2011;**1**:282-285. DOI: 10.7763/IJBBB.2011.V1.53

[45] Chai H, Yao J, Sun J, Zhang C, Liu W, Zhu M, Ceccanti B. The effect of metal oxide nanoparticles on functional bacteria and metabolic profiles in agricultural soil. Bulletin of Environmental Contamination and Toxicology. 2015;**94**(4):490-495. DOI: 10.1007/s00128-015-1485-9

[46] Chaussod R, Houot S, Guirand D, Heteir JM. Size and turnover of the microbial biomass in agricultural soils: Laboratory and field measurements. In: Jenkinson DS, Smith KA, editors. Nitrogen Efficiency in Agricultural Soils. Amsterdam: Elsevier Applied Science; 1988. pp. 312-326

[47] Jenkinson DS, Ladd JN. Microbial biomass in soil: Measurement and turnover. In: Paul EA, Ladd JN, editors. Soil Biochemistry. New York: Marcel Dekker; 1981. pp. 415-471

[48] Borneman J, Hartin RJ. PCR primers that amplify fungal rRNA genes from environmental samples. Applied and Environmental Microbiology. 2000;**66**:4356-4360

[49] Fierer N, Leff JW, Adams BJ, Nielsen UN, Bates ST, Lauber CL, Owens S, Gilbert JA, Wall DH, Caporaso JG. Cross-biome metagenomic analyses of soil microbes. Proceedings of the National Academy of Sciences. December 2012;**109**(52):21390-21395. DOI: 10.1073/pnas.1215210110

Chapter 3

Understanding Species Traits and Biodiversity Indices to Solve Problems Associated with Legume Persistence in Cropping Systems

M. Anowarul Islam and Dennis Shibonje Ashilenje

Additional information is available at the end of the chapter

http://dx.doi.org/10.5772/intechopen.76523

Abstract

Shading and competition for mineral nutrients by grass impair legume functions and production in mixed cropping systems. Sustained stress from competition and adverse environments contribute to shortened legume life spans in such cropping systems. This creates negative consequences to forage productivity. There are opportunities to solve the challenge of legume persistence by understanding species traits and plant community dynamics that foster coexistence and complementary resource use. Together with species' unique ability to tolerate adverse soil factors such as water stress, acidity and salinity, self-seeding, and shade tolerance are positive traits among legume species that grow in mixed crops. In communities, converging leaf and shoot conformations as well as asynchrony in dry matter distribution among species can avert negative effects of species competition. While seeding ratios can influence forage production and quality, management including harvest frequency and optimizing phosphorus (P) and potassium (K) fertilizers have crucial roles in perpetuating legume growth and function in mixtures with grass. Some facts on species competition for light, water, and nutrient resources; shade avoidance; and biodiversity mechanisms are highlighted in this chapter.

Keywords: legume persistence, competition, species traits, biodiversity mechanisms, crop management

1. Introduction

Legumes are important components of cropping system because of their ecosystem services. Legumes are a rich source of protein as both grains and forages. However, for decades, this group of plants has received immense attention particularly due to their unique ability to fix

IntechOpen

© 2018 The Author(s). Licensee IntechOpen. This chapter is distributed under the terms of the Creative Commons Attribution License (http://creativecommons.org/licenses/by/3.0), which permits unrestricted use, distribution, and reproduction in any medium, provided the original work is properly cited. (cc) BY atmospheric nitrogen (N). This primarily involves symbiotic relationships with N-fixing bacteria. Thus, under suitable conditions, the amount of N benefit from legumes can be enough to substitute for inorganic N fertilizers. To this effect, legumes are integrated in non-legume crops as intercrop or rotational crop [1]. On the other hand, the need for production of forages sufficient to support optimum livestock production calls for combination of grass with legume crops. The benefits of such cropping systems are well documented. This includes enhanced forage biomass production, crude protein, and digestibility [2–4].

Certain species of legumes are endowed with properties that can boost forage utility. For instance, sainfoin (*Onobrychis viciifolia* Scop.) and birdsfoot trefoil (*Lotus corniculatus* L.) contain tannins, which bind to foaming agent in alfalfa (*Medicago sativa* L.) responsible for bloat, a livestock disorder [5]. This, therefore, lauds the need for diverse array of legume species in a crop. Besides their feed value, an assemblage of legumes with varied abilities to adapt to the local environment is an ingredient for sustainable cropping systems. This is particularly important for perennial cropping systems. The characteristics of importance include tolerance to acidity, drought, shade, salinity, and heavy metals. However, such benefits are not guaranteed when legumes succumb to competition for resources key to their survival and net primary production. In this context, light, mineral nutrients, and water are few of key factors vital for legumes. In addition, temperature extremes, pests and disease damage, and herbivory are unique challenges that can reduce legume persistence. In cultivated systems, the management of mineral nutrients and irrigation, together with disease and pests control, can optimize crop growth and maintain legumes in mixtures. For some species, intra-annual persistence is influenced by harvesting practices. For example, sainfoin has poor regeneration after first harvest (**Figure 1**). In this regard, frequent and early cutting can weaken plants to the point of death. Early cutting limits the amount of food reserve required to maintain plant vigor and persistence [6].

Cold temperatures are particularly detrimental to legumes. Winter injury includes intra and intercellular freezing of unhardened plant tissues and physical damage to roots caused by ice heaving [7]. Plants exposed to freezing temperatures may also indirectly suffer from dehydration when water in plant tissues is bound in ice [8]. Frost damage causes intracellular freezing.

Figure 1. Poor growth of sainfoin in mixture with meadow bromegrass after second harvest in August, 2016. The photo was taken on October 21, 2016, at the University of Wyoming Sheridan Research and Extension Center, Wyoming, USA. Photo by D.S. Ashilenje.

Figure 2. Early symptoms of frost damage to alfalfa in a mixture with meadow bromegrass and birdsfoot trefoil. The other crops seem not to have been affected by the frost. The photo was taken on October 21, 2016 at the University of Wyoming Sheridan Research and Extension Center, Wyoming, USA. Photo by D.S. Ashilenje.

Extensive damage triggers loss of plant's photosynthetic surface, which impairs tiller development and plant growth. **Figure 2** shows early signs of frost injury to alfalfa growing in a mixture with meadow bromegrass (*Bromus biebersteinii*; Roem & Schult) and birdsfoot trefoil. It is evident that meadow bromegrass and birdsfoot trefoil have not been affected by the frost. This attests to the varying abilities among grass and legume species to tolerate excessively cold temperatures. On the other hand, winter hardiness is dependent upon hardened plant tissues and reduced leaf growth in fall. To sum up, species persistence is determined by genetic and physiological traits that enable species to acclimate to less ideal environments in multiple cropping systems.

2. Elucidation of resource competition and its influence on legumes

In mixed cropping systems, plant species compete for resources (e.g., light, water, nutrients) for their growth and survival. Competition for resources along with adverse environmental factors can negatively affect plant growth and contribute to shortened life spans of stressed plants in the cropping systems.

2.1. Competition for light

Like other plant species, legumes intercept the photosynthetic active radiation for photosynthesis [9]. This is a major input of carbon and substrates for plant's energy needs. However, for legumes, the energy gained from photosynthesis is of intermediate benefit to bacterial symbionts involved in N fixation. The daily rate of carbon assimilation by legumes growing together with grass is a hyperbolic function of leaf area index [10]. This is usually expressed as extinction coefficient abbreviated as k [11]. In practice, net accumulation rate is derived from dry matter accumulation per unit leaf area. This function is correlated to the number of leaf surfaces in the crop canopy exposed to light [12]. This partly determines shading of legumes by grasses when intercropped.

Grass leaf orientation and placement obstructs light from under-canopy growth of legumes. This phenomenon is well discussed [11]. Varying leaf angles in distinct species influence canopy extinction coefficient whereby vertically inclined or small leaves as the case is in grasses have low k values ranging between 0.3–0.5. Furthermore, leaves with clamped sheaths around the stem have intermediate *k* measurements, while species with horizontal leaves may have higher values of up to 0.7–0.8. Dense crops share space with additional vertical leaves that have low k values, that is, there are more leaves in the same area of as one horizontal leaf but with much less mutual shading.

2.2. Plant response to light quality

Plant species have profound ways of modifying growth in response to changing light quality. Annual crop species versus weed species signaling is controlled by red to far red (R:FR) ratio [13]. This interaction explains changes in plant forms in preparation for competition. High plant density absorbs incoming solar radiation causing decreased R:FR [14]. Whereas less dense plants cause an increase in R:FR ratio due to light reflection, high R:FR ratio triggers the plants to adapt to lesser light exposure due to shade from additional grass blades as enhanced growth of hypocotyl and leaf petioles/blades show (**Figure 3**) which assume an erect position [15].

2.3. Consequences of depriving legumes of light to their ecosystem services

Earlier publications have shown the adverse effects of shading on growth and dry matter production of legumes. The influence of partial (50–60%) and intense shade (80–90%) on selected cool- and warm-season legume monocrops are shown in **Table 1**. Partial shading suppresses yields for alfalfa and Illinois ticktrefoil (*Desmodium illinoense*; A. Gray) in ranges of 15–78%, respectively. Nevertheless, intense shade causes yield reductions ranging from 17% for Illinois tick trefoil to 73% for sub clover (*Trifolium subterraneum L.*).

Nitrogen fixation is also adversely affected by shading. It has been demonstrated that legumes shaded by grass have reduced size of nodules [16]. This impairs their ability to fix N. However,

Figure 3. Meadow bromegrass with elongated leaf sheaths and blades representing shade avoidance when grown in mixture with alfalfa at the University of Wyoming Sheridan Research and Extension Center, Wyoming, USA. Photo by D.S. Ashilenje.

Crop	Percent reduction in yield (% shading)		References
	Partial shading	Intense shading	
Alfalfa	15 (50)	39 (80)	[18]
Alsike clover (*Trifolium hybridum L.*)	42 (50)	68 (80)	[18]
Birdsfoot trefoil	36 (50)	69 (80)	[18]
White clover (*Trifolium repens L.*)	19(50)	41(80)	[18]
Sub clover	46 (60)	73 (90)	[19]
Illinois tick trefoil	78 (55)	17 (80)	[20]

Table 1. Yield reduction for selected legumes in response to various levels of shading.

recent evidence from modeling experiments reveals interactions that are more complicated. For instance, Schwinning and Parsons [10] have suggested that grasses not only shade, but they also take advantage of enhanced N fixation by dwindling legume densities in mixed stands. Species of legume elicit different response to shading in their N fixation [17]. This is exemplified by kudzu (*Pueraria lobata;* Ohwi), a tropical pasture legume, whose N fixation suffers less from effects of shading compared to siratro (*Macroptilium atropurpureum* cv. siratro).

2.4. Competition for soil nutrients

The antagonistic role of mineral N against biological N fixation by legumes is well known. Such effects get pronounced when N is availed in amounts exceeding that required in first few weeks of legume seedling establishment [21]. Conversely, other nutrients including P, K, molybdenum, and iron synergize de-nitrogen (N_2) fixation by legumes. Phosphorus has a direct role of promoting nitrogenase activity, the enzyme involved in conversion of N_2 to ammonia [22]. On the other hand, P increases nodule mass. Work by Mendoza et al. [23] has revealed that optimum supply of P triggers a positive growth response of legumes integrated with grass crops unlike pure stand. However, P may favor grass shoots and root biomass accumulation at the expense of the development of congregate legumes.

Potassium is essential for many metabolic processes important to plant growth. This includes photosynthesis, osmoregulation, protein synthesis, and enzyme activation [18]. More specifically, K has been correlated with increased nodulation and N_2 fixation by legumes [24]. In the same note, water stress can inhibit nodulation and nitrogenase activity. Thus, K averts injurious effects of soil water deficit on nitrogenase activity. Grass and legumes have enhanced competition for K when growing together in mixtures [25]. This phenomenon is supported by the finding that increased supply of K enhances N_2 fixation and uptake of P and N by tropical legumes [25].

3. The challenge of legume persistence in mixed stands

Legume persistence entails continuity of individual plant as well as crop stand. Legume stand persistence refers to perpetual number of individuals representing the species, which addresses the needs unique to ecosystems [26, 27]. Plant persistence is of greater concern during crop establishment after which stand persistence becomes important [26]. Efforts to improve forage crops have targeted monocrop yields, and ability to withstand factors that militate against crop survival including pests, diseases, drought stress, winter kill, soil salinity, and aluminum toxicity [27]. Previously, it has been shown that legumes persist less in mixtures because of exploitation by grasses for light and fixed N as earlier mentioned [10]. This scenario is evident from changes in morphological traits [28] as well as resource allocation [29]. Both species traits and management practices influencing forage monocrop persistence have been described in detail by Bouselinck et al. [26]. For instance, crown formers, specifically alfalfa, reach physiological potential 2 years after planting after which yields are stable up to 6–8 years of crop lifespan. In contrary, plant density declines rapidly from 300 plants m^{-2} in the seeding year to 50 plants m^{-2} in the third year attributed to intraspecific competition and disease. In addition, Louarn et al. [30] recorded a decline from 352 plants m^{-2} during first harvest to 90 plants m^{-2} in the third harvest within 1 year. Declining number of alfalfa plants is compensated for by increasing number of stems $plant^{-1}$. For self-seeding species, notably birdsfoot trefoil, persistence depends on their ability to set enough seeds, thereby building substantial seed banks [31].

4. Biodiversity indices explain complex species interactions in polycultures

Various mechanisms underlie biodiversity in forage plant communities. Among these are those that enhance temporal stability. Temporal stability refers to constancy in species abundance [32]. In this case, abundance may be determined as biomass production or density. Temporal stability is computed from the inverse of the coefficient of variation or ratio between mean abundance to corresponding standard deviation. Community temporal stability derives from lower variance in averaged production of many species in a community than individual species, which is also referred to as portfolio effect [32].

According to Isbell et al. [33], a positive correlation exists between temporal stability and other biodiversity indices, namely overyielding and species asynchrony. Overyielding refers to higher biomass production for a mixed crop when compared to the average pure stand of the species constituted in the mixture [34]. Positive species interactions promote greater yields for a mixture when compared to best performing species in the mixture [35]. However, the overyielding effect is often diluted by the role of dominant species in equilibrating biomass production for mixtures with that of its pure stand which has been extensively discussed [34, 36]. On the other hand, Isbell et al. [33] have shown that no relation existed between temporal stability and species evenness. They explain this by the asynchronized dominance in biomass production by distinct species over a time scale.

5. Plant species traits responsible for species persistence in mixed stands

Studies concerning biodiversity mechanisms suggest possible ways to perpetuate legumes in mixed crops. This is typified by asynchronous dry matter distribution [33] and convergence in species traits [37]. This property of plant communities permits complementary use of light, water, and nutrients. Such species characteristics include specific leaf area (SLA) which is the ratio of leaf area to corresponding dry weight. The SLA is an adaptation to relatively larger leaf surface compared to leaf carbohydrate reserve. Thus, SLA is correlated with greater photosynthetic capacity and leaf N concentration. Conversely, low SLA has been linked to longer leaf lifespan and retention of nutrients [9]. On the other hand, Gubsch et al. [29] have explained the diversity and adjustment of functional traits to environment as a determinant of forage production. For instance, grass species adjustment to low light intensity and improved N acquisition contributes to increasing forage quality as more species are included in a mixture. Besides, divergent leaf forms and convergent plant configurations, such as shoot height, to a greater extent, account for legume persistence [37].

As mentioned in the introduction, storage of food reserves can spur legume persistence. The daily rate of carbon assimilation by legumes and grasses growing together is determined by light interception [10]. Light interception by each species is a function of the quantity of radiant energy received at a surface per unit time (photosynthetic photon flux density or PPFD), leaf area index (LAI), and canopy extinction coefficient [11]. Light interception, carbon assimilation, and plasticity in plant morphology are dynamic interactions that impinge on overall plant development and persistence. For instance, shading has been found to reduce secondary branching and plant leaf area development in dense crops [38]. Whereas in severe competition, growth of primary axis is impaired, different plant configurations influence forage nutritive value notably crude protein (CP) and acid detergent fiber (ADF). For example, alfalfa CP is positively correlated to the ratio between leaf and stem weight but negatively correlated to stem length and maturity [39]. The relation is inversely proportional to ADF. In grazed systems, higher leaf-to-stem ratio toward the top of the canopy correlates with higher CP. Thus, grazing animals selectively graze on the apical regions of the legume canopy.

6. Lessons learned from recent studies involving tall fescue-alfalfa mixtures

Competitiveness among species in mixtures was earlier quantified as relative yield total [40]. Relative yield total for combined species is computed from summing up the product of dry matter proportion and ratios of mixture versus monocrop yields. Relative yield total values greater than 1 indicate complementarity, while values less than 1 show competition among species for available resources. Relative yield total is dependent upon management and environment [41]. The other measure of competitiveness is species aggressivity abbreviated as A. As defined by [42] when species aggressivity for one taxon is >1, then that species is more competitive than its contemporary growing in the mixture.

Results from a recent experiment at the James C. Hageman Sustainable Agriculture Research and Extension Center, Lingle, Wyoming, USA, are presented in the rest of this section. There was a significant (P = 0.002) interaction between the year and tall fescue (TF) (*Schedonorus arundinaceus* [Schreb.] Dumort.)—alfalfa seeding ratios to influence proportion of alfalfa in mixtures (**Table 2**). The 25:75% mixture of tall fescue with alfalfa maintained proportions of legume biomass at a minimum of 64% from 2012 to 2015. However, in the same duration, there were slight fluctuations in the proportions of alfalfa biomass in the 75:25% mixture of tall fescue alfalfa (56–47) and 75:25% mixture of tall fescue alfalfa (43–64). The proportion of alfalfa biomass in the 25:75% mixture of tall fescue alfalfa only surpassed that of the 75:25% mixture of tall fescue alfalfa in the years 2013 and 2014. In contrary, diverse mixtures of tall fescue and alfalfa did not influence ($P = 0.22$) relative yield total. In this regard, all mixtures of tall fescue and grass had relative yield totals >1 (**Table 3**). Similar results were recorded for grass aggressiveness which did not vary significantly ($P = 0.218$) across treatments. On the other hand, varying tall fescue—alfalfa seeding ratios did not affect aggressiveness of tall fescue in mixtures. Harvest frequency interacted with year ($P < 0.0001$) to influence species competition. Except for the year 2015, early growth during spring had relative yield total values <1 which depicts competitive growth among species (**Figure 4**). However, increased harvesting frequency gave relative yield total values >1 averaged across different seeding ratios. Between the years 2012 and 2014, tall fescue had more aggressive growth after the first harvest (**Figure 5**). However, in 2015, there was no competitive advantage in growth of tall

	Proportion of alfalfa biomass			
	Year			
Treatment	**2012**	**2013**	**2014**	**2015**
	%			
TF:alfalfa 75:25 seeding ratio	56aA[†]	48aA	54abA	47aA
TF:alfalfa 50:50 seeding ratio	43bA	47aA	44aA	61bB
TF:alfalfa 25:75 seeding ratio	64aA	67bA	64bA	64bA

[†]Within column refers to different treatments followed by same letter in lower case and within row refers to different years followed by same letter in upper case are not significantly different at $P < 0.05$

Table 2. Proportions of alfalfa biomass in tall fescue (TF) and alfalfa mixed crops established using different seeding ratios at James C. Hageman sustainable agriculture research and extension center near Lingle, Wyoming, USA, during 2012–2015.

Grass legume mixture	**Relative yield total**
TF:alfalfa 75:25 seeding ratio	2.07
TF:alfalfa 50:50 seeding ratio	1.99
TF:alfalfa 25:75 seeding ratio	1.60
Mean	1.89
P-value	0.22
LSD (0.05)	0.59

Table 3. Species relative yield total for tall fescue (TF) and alfalfa mixed crops established using different seeding ratios at James C. Hageman Sustainable Agriculture Research and Extension Center near Lingle, Wyoming, USA.

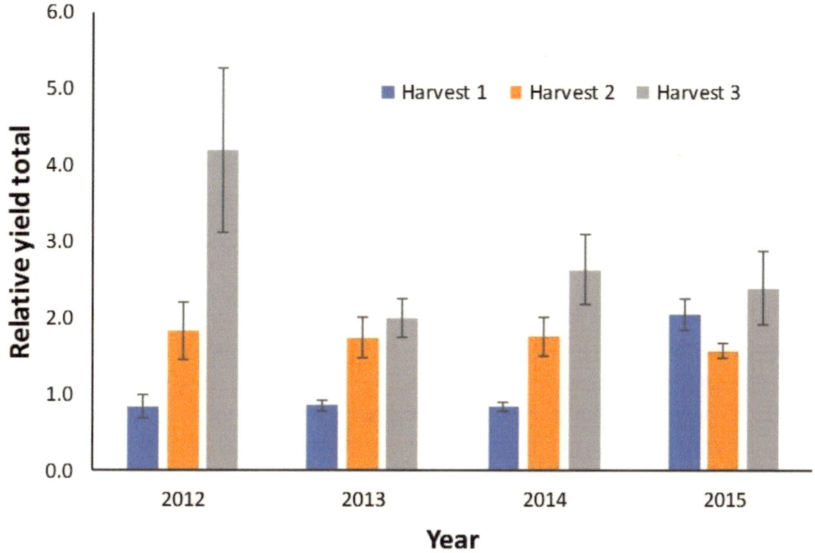

Figure 4. Mean relative yield total across different cropping mixtures of tall fescue and alfalfa observed at different harvest times during 2012–2015 at James C. Hageman Sustainable Agriculture Research and Extension Center near Lingle, Wyoming, USA. The mixtures were tall fescue-alfalfa in 75:25, 50:50, and 25:75 seeding ratios. Bars represent standard error of the mean.

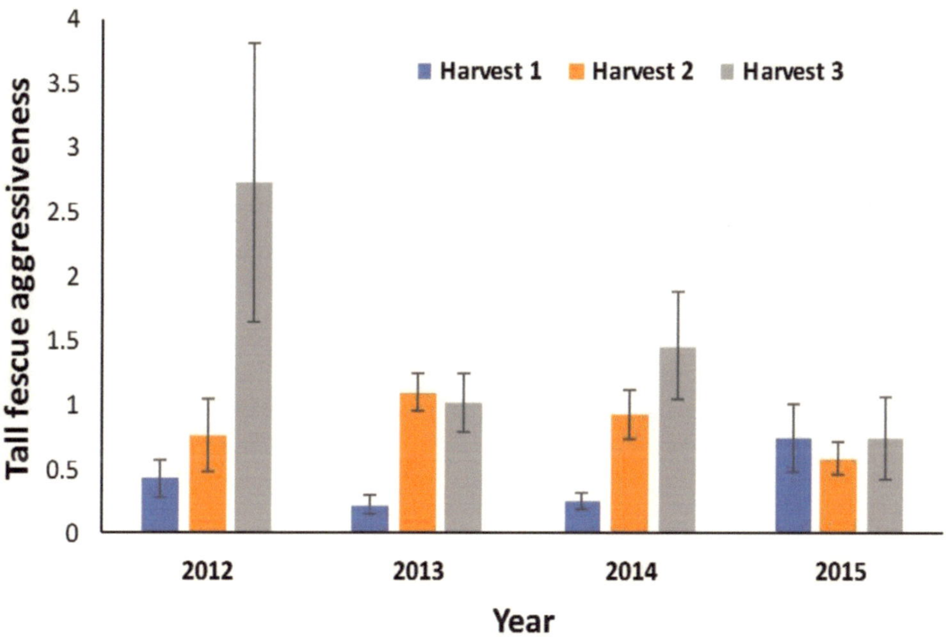

Figure 5. Tall fescue aggressiveness observed at different harvest times during 2012–2015. The values were averaged across different mixtures of tall fescue-alfalfa in 75:25, 50:50, and 25:75 seeding ratios at James C. Hageman Sustainable Agriculture Research and Extension Center near Lingle, Wyoming, USA. Bars represent standard error of the mean.

fescue when compared to alfalfa. This was same across different crop harvests. These results suggest that there was complementarity in resource use regardless of the seeding ratios for tall fescue-alfalfa mixtures. Therefore, this sustained legume growth despite dominance of grass in the mixtures.

7. Conclusions

Legume crops are treated as natural fertilizer because of their unique ability to fix atmospheric N. Considering their role in N_2 fixation and rich nutritive value, legumes are valuable inclusions in forage cropping systems with grasses. However, the realization of these benefits is limited because of poor legume persistence or disappearance of legumes in mixed cropping systems. Grass species have more aggressive growth, thus they compete against legumes for light, water, and nutrients. Such competition can impair N_2 fixation and growth and finally lessen legume persistence. Knowledge about species tolerance for shade and biodiversity mechanisms can help forestall belligerence from grasses and perpetuate legumes in mixed cropping systems. Embedded in legume persistence is diverse species trait that contributes to asynchronized growth patterns and leaf forms. Therefore, the species traits and biodiversity indices that can help solve the problems of legume persistence are the focus in this chapter. Several breakthroughs that emerged from the review and recent experiments include:

- Genetic and physiological traits enable species to acclimate to less ideal environments in multiple cropping systems including temperature extremes, drought, and shading.
- Species may display hardening of plant tissues to resist freezing temperatures and leaf forms and placement that facilitates more assimilation of light to form food reserves.
- Biodiversity mechanisms that encourage complementary use of resources can help to alleviate loss of legumes in mixed cropping systems leading to overyielding.
- Crop management, particularly harvesting frequency, plays more significant role than seeding proportions in influencing competitiveness of grass against their companion legumes in the mixtures.
- Finally, despite the aggressive growth of associated species in mixed cropping systems, complementary resource use allowed legumes to thrive 4 years or more after their establishment in different mixtures.

Author details

M. Anowarul Islam* and Dennis Shibonje Ashilenje

*Address all correspondence to: mislam@uwyo.edu

Department of Plant Sciences, University of Wyoming, Laramie, WY, USA

References

[1] Stagnari F, Maggio A, Galieni A, Pisante M. Multiple benefits of legumes for agriculture sustainability: An overview. Chemical and Biological Technologies in Agriculture. 2017;**4**: 2-13. DOI: 10.1186/s40538-016-0085-1

[2] Berdahl JD, Karn JF, Hendrickson JR. Dry matter yields of cool-season grass monocultures and grass alfalfa binary mixtures. Agronomy Journal. 2001;**93**:463-467

[3] Sanderson MA, Brink G, Stout R, Ruth L. Grass-legume proportions in forage seed mixtures and effects on herbage yield and weed abundance. Agronomy Journal. 2013;**105**: 1289-1297. DOI: 10.2134/agronj2013.0131

[4] Adjesiwor AT, Islam MA, Zheljazkov VD, Ritten JP. Garcia y Garcia A. Grass-legume mass seed mass ratios and nitrogen rates affect forage accumulation, nutritive value and profitability. Crop Science. 2017;**57**:2852-2864. DOI: 10:2135/CROPSCI2016.09.0776

[5] Wang Y, McAllister TA, Acharya S. Condensed tannins in sainfoin: Composition, concentration, and their effects on nutritive and feeding value of sainfoin forage. Crop Science. 2015;**55**:13-22. DOI: 10.2135/cropsci2014.07.0489

[6] Smith D. Physiological considerations in forage management. In: Heath ME, Metcalfe DS, Barnes RF, editors. Forages: The Science of Grassland Agriculture. 2nd ed. Iowa, USA: The Iowa State University Press; 1973. p. 431

[7] Larsen A. Breeding winter hardy grasses. Euphytica. 1994;**77**:231-237. DOI: 10.1007/BF02262635

[8] Volenec J, Nelson CJ. Environmental aspects of forage management. In: Barnes RF, Nelson CJ, Collins M, Moore KJ, editors. Forages: An Introduction to Grassland Agriculture. 6th ed. New York, USA: Blackwell Publishing; 2003. pp. 108-117

[9] Lambers H, Chapin FT III, Pons TL. Plant Physiological Ecology. 2nd ed. New York, USA: Springer; 2008. p. 16. DOI: 10.1007/978-0-387-78341-3

[10] Schwinning S, Parsons AJ. Analysis of the coexistence mechanisms for grasses and legumes in grazing systems. Journal of Ecology. 1996;**84**:799-813

[11] Chapin FS, Matson PA, Vitousek PM. Principles of Perennial Ecosystem. 2nd ed. New York, USA: Springer; 2011. p. 140. DOI: 10.1007/978-1-4419-9504-9

[12] Tesar MB. Physiological basis of crop growth and development. American Society of Agronomy, Crop Science Society of America. Madison, Wisconsin; 1984. pp. 157-160. DOI: 10.2135/1984.physiologicalbasis.frontmatter

[13] Page ER, Tollernaar M, Lee EA, Lukensand L, Swanton CJ. Shade avoidance: An integral component of crop-weed competition. Weed Research. 2010;**50**:281-288

[14] Aphalo PJ, Ballare CL, Scopel AL. Plant-plant signaling the shade-avoidance response and competition. Journal of Experimental Botany. 1999;**50**:1629-1634

[15] Casal JJ. Shade avoidance. The Arabidopsis Book. American Society of Plant Biologists. 2012;**10**. DOI: 10.1199/tab.0157

[16] Chu ACP, Robertson AG. The effect of shading and defoliation on nodulation and nitrogen fixation by white clover. Plant and Soil. 1974;**41**:509-519

[17] Fujita K, Matsumoto K, Ofosu-Budu GK, Ogata S. Effect of shading on growth and dinitrogen fixation of kudzu and tropical pasture legumes. Soil Science & Plant Nutrition. 1993;**39**:43-54

[18] Lin CH, McGraw RL, George MF, Garrett HE. Shade effects on forage crops with potential in temperate agroforestry practices. Agroforestry Systems. 1999;**44**:109-119

[19] Kyriazopoulos AP, Abraham EM, Parassi ZM, Koukoura Z, Nastis AS. Forage production and nutritive value of *Dactylis glomerata* and *Trifolium subterraneum* mixtures under different shading treatments. Grass and Forage Science. 2012;**68**:72-82

[20] Houx III JH, Mc Graw RL, Fritschi FB, Navarrete-Tindall NE. Effects of shade on growth and nodulation of three native legumes with potential for use in agroforestry. Native Plants Journal. 2009;**10**:232-238

[21] Marschner H. Mineral Nutrition of Higher Plants. 2nd ed. California, USA: Academic Press; 1995. pp. 229-404. DOI: 10.1016/B978-0-08-057187-4.50001-1

[22] Olivera M, Tejera N, Iribarne C, Ocaña A, Lluch C. Effect of phosphorus on nodulation and nitrogen fixation by *Phaseolus vulgaris*. In: Velãzquez E, Rodríguez-Barrueco C, editors. First International Meeting on Microbial Phosphate Solubilization. Development in Plant and Soil Sciences. Vol. 102. Dordrecht: Springer; 2007. pp. 157-160. DOI: 10.1007/978-1-4020-5765-6_23

[23] Mendoza R, García I, Depalma D, Lõpez CF. Competition and growth of a grass-legume mixture fertilized with nitrogen and phosphorus: Effect on nutrient acquisition, root morphology and symbiosis with soil microorganisms. Crop & Pasture Science. 2016;**67**: 629-640

[24] Wahab AM, Abd-Alla MH. The role of potassium fertilizer in nodulation and nitrogen fixation of faba bean (*Vicia faba* L.) plants under drought stress. Biology and Fertility of Soils. 1995;**20**:147-150

[25] Nurjaya IGMO, Tow PG. Genotype and environmental adaptations as regulators of competitiveness. In: Tow PG, Lazenby A, editors. Competition and Succession in Pastures. London, UK: CABI Publishing; 2001. p. 56

[26] Bouselinck PR, Bouton JH, Lamp WO, Matches AG, McCaslin MH, Nelson CJ, Rhodes LH, Sheaffer CC, Volenec JJ. Legume persistence in forage crop systems. Journal of Production Agriculture. 1994;**7**:311-322

[27] Bouton JH. Breeding Lucerne for persistence. Crop & Pasture Science. 2012;**63**:95-106. DOI: 10.1071/CP12009

[28] Riday H, Brummer EC. Vigor and persistence of birdsfoot trefoil populations selected with or without an orchardgrass companion evaluated in grass sod. Crop Science. 2014; **54**:2010-2076. DOI: 10.2135/cropsci2014.02.0147

[29] Gubsch M, Buchmann N, Schmid B, Schulze E, Lipowsky A, Roscher C. Differential effects of plant diversity on functional trait variation of grass species. Annals of Botany. 2011;**107**:157-169

[30] Louarn G, Pereira-Lopes E, Fustec J, Mary B, Voisin A, Carvalho PC, Gastal F. The amount and dynamics of nitrogen transfer to grasses differ in alfalfa and white clover-based grass-legume mixtures as a result of rooting strategies and rhizodeposit quality. Plant and Soil. 2015;**389**:289-305

[31] Ayres JF, Kelman WM, Wiedman SG, McCorkell BE. Developing birdsfoot trefoil (*Lotus corniculatus* L.) varieties for permanent pasture applications in low latitude regions of eastern Australia. Australian Journal of Experimental Agriculture. 2008;**48**:488-498. DOI: 10.1071CEAO7114

[32] Tilman D. The ecological consequences of changes in biodiversity: A search for general principles. Ecology. 1999;**80**:1455-1474. DOI: 10.1890/00129658(1999)080[1455:TECOC1] 2.0.CO

[33] Isbell FI, Polley HW, Wilsey BJ. Biodiversity, productivity and temporal stability of productivity patterns and processes. Ecology Letters. 2009;**12**:443-445. DOI: 10.1111/j.1461-0248.2009.01299.x

[34] Kirwan L, Lüscher A, Sebastai MT, Finn JA, Collins RP, Porqueddu C, Helgadottir A, Delgado I, Elgersma A, Fothergill M, Frankow-lindberg BE, Golinski P, Grieu P, Gustavason AM, Kadziuliene Z, Kanjoris T, Lunnan T, Malengier M, Maltoni S, Meyer V, Nyfeller D, Nykanen-kurki P, Parente J, Smit HJ, Thumm U, Connolly J. Evenness drives consistent diversity effects in intensive grassland systems across 28 European sites. Journal of Ecology. 2007;**95**:530-539. DOI: 10.1111/j.1365-2745.2007.01225.x

[35] Bonin CL, Tracy BF. Diversity influences forage yield and stability in perennial prairie plant mixtures. Agriculture, Ecosystems and Environment. 2012;**162**:1-7

[36] Schmid B, Hector A, Saha P, Loreau M. Biodiversity effects and transgressive overyielding. Journal of Plant Ecology. 2008;**1**:95-102

[37] Pellkofer S, van der Heijen MGA, Schmid B, Wagg C. Soil communities promote temporal stability and species asynchrony in experimental grassland communities. PLoS One. 2016; **11**. DOI: 10.1371/journal.pone.0148015

[38] Baldissera TC, Frak E, Carvalho PC, Louarn G. Plant development controls leaf expansion in alfalfa plants competing for light. Annals of Botany. 2014;**113**:145-157. DOI: 10.1093/a0b/met251

[39] Hakl J, Fuksa P, Konecna J, Santrucek J. Differences in the crude protein fractions of Lucerne leaves and stems under different stand structures. Grass and Forage Science. 2016;**71**:413-423

[40] de Wit CT, Van den Berg JP. Competition between herbage plants. Netherlands Journal of Agricultural Science 1965;**96**:561-567

[41] Connor DJ, Loomis RS, Cassman KG. Crop Ecology: Productivity and Management in Agricultural Systems. 2nd ed. New York, USA: Cambridge University Press; 2011. p. 61

[42] Hoffman ML, Buhler DD. Utilizing Sorghum as a functional model of crop-weed competition. Establishing a competitive hierarchy. Weed Science. 2002;**50**:466-472

Section 2

Mixed Cropping Systems and Plant Competition

Chapter 4

Competitive Ability of Rice Cultivars in the Era of Weed Resistance

Fábio Schreiber, Ananda Scherner, André Andres, Germani Concenço and Francisco Goulart

Additional information is available at the end of the chapter

http://dx.doi.org/10.5772/intechopen.78342

Abstract

Almost all plants are negatively affected by neighboring plants, which impose some degree of competition within the population, depending mainly on the quantity and quality of natural resources available in the environment. In rice cultivation, the occurrence of a high and diverse infestation of weeds results in high competition levels among the species. In addition, the high and growing number of cases about herbicide-resistant weeds, especially the widespread distribution of Imidazolinone-resistant weedy-rice and the high infestation of weeds belonging to the *Echinochloa* genus, has increased the competition levels within rice cultivation due to the lack of control. Therefore, the inclusion of rice cultivars with greater competitive ability represents a promising tool for weed management, since new cases of resistance to herbicides are often reported and alternative control strategies are scarce. The use of rice cultivars with a greater ability to suppress weeds can alleviate the competitive effect of these species, giving priority to the crop for the use of environmental resources due to the faster occupation of the ecological niches. Thus, this chapter aims to explore the competitive ability of rice cultivars against troublesome weed species, accounting for the role of their morphological and physiological traits as a function of environment-friendly crop practices.

Keywords: weed, competition, weed free period, competitiveness traits

1. Introduction

The occurrence of a high and diverse weed infestation in paddy rice is among the various adversities that can be encountered during the crop life cycle, which might hamper crop yields. Weeds compete with the crops for natural resources that enable them to survive and

IntechOpen

© 2018 The Author(s). Licensee IntechOpen. This chapter is distributed under the terms of the Creative Commons Attribution License (http://creativecommons.org/licenses/by/3.0), which permits unrestricted use, distribution, and reproduction in any medium, provided the original work is properly cited.

reproduce, such as light, water and nutrients. Thus, the presence of a very diverse weed community within a field, together with the high rate of occurrence, makes the control difficult, and has negative consequences on rice grain quantity and quality increasing as well the production costs [1].

Competition can be defined as an interaction between individuals or populations, which is negative for both and it is rare to find a plant which has not been affected by neighboring plants [2]. Within a plants' community, competition is generally indirect, in which one individual affects negatively another by taking up resources that are limited in the environment, and could otherwise be available for other individuals. Direct competition can occur within a plant community, but these mechanisms are rare or still unknown, such as allelopathy. In general terms, it is very difficult to establish the cause for competition, because in natural systems, multiple resources are often simultaneously limited [3].

The losses in rice yield due to weed competition vary with the system of crop implantation: conventional system, minimum tillage, no-tillage, pre-germinate, pre-germinate mix, and transplant seedlings; with rice cultivars (e.g., cycle and height) with soil fertility; with the weeds present in the crop (e.g., species, density, duration and time of occurrence); and with management practices [4]. In areas where weed control strategies are not applied, the reduction in productivity can reach almost totality. Significant reductions in world rice production are estimated at 35, 24 and 16%, respectively, due to weeds, pests and pathogens [5]. In lowland rice crop, in Brazil, there were decreases in production caused by weeds from 50 to 100%. Therefore, this crop is quite sensitive to weed interference [4].

The most troublesome weed species that occur in Brazilian paddy rice are *Oryza sativa* (weedy-rice), *Echinochloa* sp. (*E. crus-galli, E. crus-pavonis* and *E. colona), Eleusine indica, Cyperus* spp. (*C. rotundus* and *C. difformis*) and *Sagittaria* spp. (*S. montevidensis*), with cases of herbicide resistant biotypes being reported for all these species. These are mainly associated with the intensive use of ALS-inhibiting herbicides, poor crop rotation schemes and crop varieties with low competitive ability. The current resistance problem demonstrates the urgent need of alternative management strategies to efficiently control these species and reduce the reliance on the chemical control. Thus, herbicide resistance and the lack of control alternatives led to the search for more competitive rice cultivars as a weed management tool in the crop technology [6].

The introduction of weed-competitive rice cultivars represents a low-cost and safe nonchemical addition to an integrated weed management (IWM) program. In addition, the use of more competitive cultivars can minimize yield losses and herbicide dependence, because these cultivars can suppress weed seed production, limit future weed infestations and fit easily into current agronomic practices [7]. However, trade-offs between competitiveness and productivity and inconsistent trait expression under weedy and weed-free conditions could complicate the breeding of competitive rice cultivars.

Crop competitiveness is a complex attribute that involves the ability to sustain yields despite the presence of weeds and the ability to suppress weed growth [8]. Thus, the competitive ability of different rice cultivars can be compared by assessing the competitive effect of plants or the ability to suppress other individuals or by assessing the competitive response of plants or the ability to avoid being suppressed.

More recently, the ecology and physiology of crops and weed species gained increasing importance in the development of methods of weed control [9, 10]. Ecology may be roughly divided in two sub-sections: synecology and autecology [11]. These areas present complementary aims in the study of ecology. Summarizing the concepts, autecology considers the species as an ecological unit while synecology considers the community as an ecological unit [11]. On the present chapter, we will focus on the competitive aspects of the autecology of rice plants, including the morphophysiological traits that confer superior competitive ability to rice, and on the phytosociological aspects of the weed communities into rice fields, e.g., the weed species against whom rice has to withstand and outstand competition.

2. Competition between plant species (interspecific competition)

Among several interpretations, "plant competition" essentially means a reduction in performance of a given plant species of importance, due to the shared use of a limited available resource [9]. Unlike animals, plants have limited mobility and, therefore, the competition between them is different, being apparently more passive and not visible at the beginning of the development [12]. However, it is known that crops, in general, do not show high competitive ability against weed species, which is the result of breeding for cultivars with productive traits and not to endure stress or aggressiveness [13].

It is of common consensus between researchers that competition occurs when neighboring plants use the same resources, and, therefore, the plant with the capacity to capture faster these resources is often more successful [13]. This capacity is normally associated with high relative growth rate, which enables the plant to capture the resources quickly, but these plants should also use the resources very efficiently. Nevertheless, it is also believed that a good competitor has both the ability to extract scarce resources and to tolerate the lack of them [13]. Thus, following this theory, a good competitor should be the species that requires fewer resources to survive, develop and reproduce [10].

In a cropping field, several weeds species can grow together with the crop cultivar in the same area. It is known that crops and the weed community tend to require similar environmental resources to survive, such as water, light, nutrients and CO_2. However, different species need these resources at different levels, but usually they are not enough even for the crop and, thus, the competition occurs. Under this situation, any plant that emerges in the cropping field will fight for these limited resources causing a reduction in crop productivity and probably reducing the quality of the harvested product as well [12].

The environmental factors that determine plant growth are commonly classified as "resources" and "conditions" [10]. Resources are the factors that can be consumed by plants such as water, CO_2, nutrients and light. Plants usually respond to resources following a standard curve, meaning that they tend to be small, if the resources are limited, and reach maximum development at the saturation point. After saturation point plant development can decline, if the resource becomes toxic (e.g., toxicity due to excessive zinc availability in the soil and water flood). On the other hand, conditions are factors not directly consumed, such as pH and soil density, which interfere in the development of plants because they can be associated reduce resources availability or plants capacity to explore them. It is always important to highlight that plant competition only occurs when the demand of a certain resource by a plant community is higher than its availability in the environment [12].

When weeds are established in the cropping fields before the crop, the competition tends to be critical and crop plants are normally inclined to fail under these circumstances [10]. However, if the crop plants are established first in the area and have similar competitive ability to the weeds, they will cover the soil reducing the weeds' access to essential resources for plant establishment such as light [9, 12, 13].

Moreover, competition is not only established between different species (interspecific competition) but can also occur among individuals of the same species (intraspecific competition). Even so, different parts of the same plant, such as leaves and roots, can compete for photo-assimilates. Based on the abovementioned aspects the following premises should be considered for the competition between crops and weeds [13]:

- Early states of crop development; the first 8 weeks for annual crops are critical for competition and this is the period where crop plants should grow free of weeds;
- Weed species that share similar biological and morphological traits with the crop are usually the most competitive when compared to those that differ greatly from crop plants;
- The size of the weeds' community is not the most important factor in terms of competition, a discreet weed infestation can be as harmful as a heavy infestation depending on the crop development stage before completion occurs;
- Direct competition between crops and weeds is established for limited environmental factors (water, light, CO_2, nutrients and physical space). Indirect competition occurs when weeds or crops release allelopathic compounds in the soil and/or air, which are capable of inhibiting the germination and/or growth of other plant species.

3. Rice traits for weed competitiveness

Crop-weed competition studies are often found in the literature and the outcome of these are normally applied when planning integrated management practices that include crop rotation with winter crops that are capable to suppress weeds and can also be used for crop-livestock integration [14, 15]. The main outcome of these studies is to model weed dynamics in the cropping fields based on their biological and morphological traits to optimize management strategies. Germination and emergence patterns, dry mass, dry mass accumulation, plants height, number of tillers or branches, number of inflorescences and other variables are often measured for future estimations [16–18].

Several traits have been associated with irrigated rice competitiveness with weeds in previous studies. Some authors believe that there is a negative correlation between competitiveness and productivity [19, 20], while others have suggested that is possible to enhance rice competitiveness and maintain high yields at the same time [7, 21]. The reasons for these divergences have not been totally elucidated, since most of these studies are based only on analyses of simple correlations and lack a mechanistic analysis of the relationships between plant characteristics that determine competitiveness and those that determine yielding ability [22, 23].

3.1. Morphological traits

Studies report the relationship between competitive morphological characteristics of cultivated plants, which offers a competitive advantage against weeds [22, 24]. Some of these characteristics are germination, growth velocity, height, canopy architecture, high biomass, leaf area and photo-assimilates [23]. For instance, rice traits associated to light capture are plant height, tillering ability, leaf morphology and area, while the development of the root system is important in terms of nutrients capture.

A number of comparative studies have shown that plant size, accounting specially for shoot length, is the main indicator of competitive ability [25]. It is clear that big plants can win over little plants, but it remains unclear whether large size confers enhanced competitive ability by reducing the resources availability to another individual or the bigger plant can tolerate reduced levels of the respective resource [2]. For instance, when light competition is in place between plants, several factors determine the ability to capture or inhibit the availability of this resource to other individuals, such as the position of the leaves. However, it is still not clear whether the leaves that have higher positions in the canopy are tolerant to lower light levels.

Adjustments in root and shoot growth are often associated with plant's phenotypic plasticity in response to changes in the environment. However, it is important to mention that not all adjustments that occur in plants size or growth rate are necessarily adaptive responses to compensate for resource limitations or competition imposed by its neighbors [26]. Some authors believe that shortages of nutrients or water could maximize a plant's probability of capturing those resources, especially if a competitor fails to respond to a comparable extent. Therefore, such responses will be associated ultimately with increased fitness and not necessarily with greater competitive ability [27]. However, the occupation of space below-ground is a fundamental characteristic of competitive success, since the nutrient uptake at the first development stages for certain species reduces the nutrients availability to neighboring plants, which indicates a competitive advantage [28].

Moreover, as plant height increases, more energy is invested for biomass production in the stem to support their own weight, which in turn reduces the fraction of leaf mass in the plant and can reflect in reduced crop yields [29]. In addition, being tall can lead to some disadvantages because these plants may be exposed to stronger winds than the neighbors, which might entail negative effects on plant growth due to excessive transpiration and mechanical stress [30, 31].

3.2. Physiological traits

Nowadays, physiological and highly specialized studies dealing with crop-weed competition still lack perspectives that could be integrated for practical everyday weed management.

Consequently, weed biologists tend to avoid the use of physiological parameters in association to the directly measured variables to support their findings. However, changes have been proposed in this scenario with the introduction of more basic research in applied studies [9, 10] propose changes to this scenario.

When rice is subjected to strong competition with weeds during cultivation, its physiological characteristics of growth and development are usually changed. This results in differences regarding the use of environmental resources, especially water, which directly affects the availability of CO_2 in leaf mesophyll and leaf temperature, therefore, the photosynthetic efficiency [32].

3.2.1. Competition for light

For some authors, competition for light is not as important as competition for water and nutrients mainly because the understanding of plant physiology traits is only starting to be included in weed studies [33]. However, it should be considered that there is an interrelation among these factors [13].

It is known that when crop plants shade completely the soil surface, there is no competition for light. Moreover, as a consequence of genetic improvement of crop cultivars, these plants tend to be more efficient in intercepting light, thus plants of crop species present high Light Use Efficiency (LUE) when evaluated alone [12]. This is probably the reason why light competition is not often included in crop-weed competition experiments. However, some studies can be found such as the one evaluating the LUE between bean and soybean crops with three weed species (*Euphorbia heterophylla, Bidens pilosa* and *Desmodium tortuosum).* The results show that crops accumulated more dry mass per unit of intercepted light than any of the studied weeds [34], but even though weeds were less efficient than crops in using light, they present high competitive ability in field conditions due to a more efficient extraction and use of other resources, like water and nutrients.

Light competition is complex because it is a result of several factors, mainly the species in question. For instance, species characteristics such as carbon metabolism of the C3, C4 or CAM types and the natural habitat (native to shaded or sunny environments) are highly important when studying light competition and will regulate the reactions that take place at the dark phase of photosynthesis [9, 12].

It is common to imagine that C4 plants are always more efficient than C3 plants; however, this is true only under certain conditions [13]. C4 plants demand more energy to produce photo-assimilates, due to the presence of two carboxylative systems. Moreover, the relation of CO_2 fixed/ATP/NADPH is 1:3:2 for C3 species and 1:5:2 for C4 species, which also evidences the higher need of energy for photosynthesis in C4 plants. Thus, it is reasonable to conclude that when the access to light is limited, C4 plants have a reduced competitive ability than C3 species because all the energy comes from light.

On the other hand, in C4 species, the enzyme responsible for carboxylation has high affinity for CO_2, which confers a high competitive ability to these species under high temperatures, light availability and also under temporary water deficit. In these situations, C4 species are capable to overcome C3 species, accumulating twice the dry mass per unit of leaf area in the same time interval [13].

3.2.2. *Competition for water*

There are various factors influencing water competition, such as the volume of soil that is covered by the rooting systems, physiological traits of the plant, stomatal regulation, osmotic adjustment in roots and hydraulic conductivity capacity of the roots [12]. Crop cultivars are normally less tolerant to water deficit than weeds, and it is common to observe crop plants with some degree of wilting, while weed plants are still completely turgid. Moreover, the competition for water is commonly associated with the competition also for light and nutrients [13].

Plant species vary in the amount of water needed per unit of dry mass accumulated; the species that use more efficiently the water are known to have high water use efficiency (WUE = amount of dry mass accumulated as a function of water used at the same period). Thus, it is reasonable to expect that species with higher WUE should be more competitive under water deficit and, therefore, more productive [10]. However, some weed species may present distinct values of WUE throughout the cycle, being more competitive for water in certain stages of their development [13].

It is very important to know the WUE of the different species within an area, although this only one of the mechanism allowing that confers water competition. In this sense, stomatal self-regulation becomes very important to overcome periods of water deficit.

3.2.3. *Competition for CO_2*

CO_2 competition is not often considered in crop-weed competition studies, because the availability of this gas is normally not considered an issue. However, it is known that plants differ in their carbon cycling mechanisms (C3 and C4 plants), resulting in different dry mass accumulation. Thus, the ability to capture CO_2 from the air is important in terms of competition because this regulates the photosynthesis under competing situation and may affect mainly C3 species [13].

3.2.4. *Case study: influence of barnyardgrass on rice physiology*

The main form of interference between barnyardgrass and irrigated rice is the competition for light and nutrients, constituting one of the main limiting factors of productivity in irrigated rice [30, 31]. In addition, it is important to note that weed competition can affect crop production and its quality, since it modifies the efficiency of use of environmental resources [35, 36].

In a study focusing on competition of Quinclorac-resistant barnyardgrass with rice plants by the additive experimental model (**Figure 1**), there were practically no differences in the accumulation of dry mass and photosynthesis, and weak differences regarding water use efficiency of rice plants as a function of competition with distinct biotypes, although rice was clearly affected by the increase in competition (**Figure 2**).

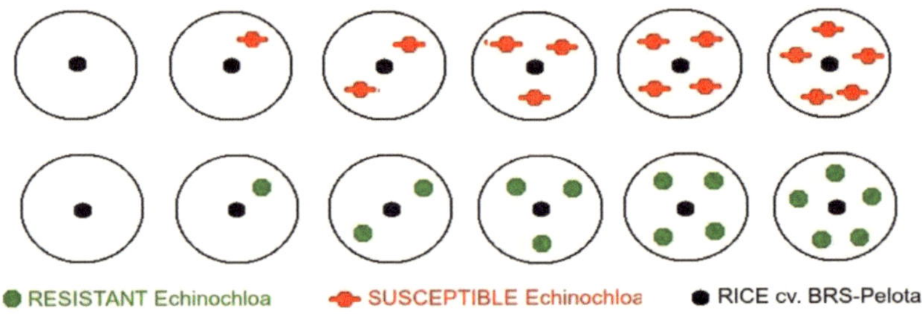

Figure 1. Schematics of the additive-model trial about the influence of barnyardgrass biotypes resistant or susceptible to the Quinclorac herbicide on the rice variety BRS Pelota, under distinct competition levels [37].

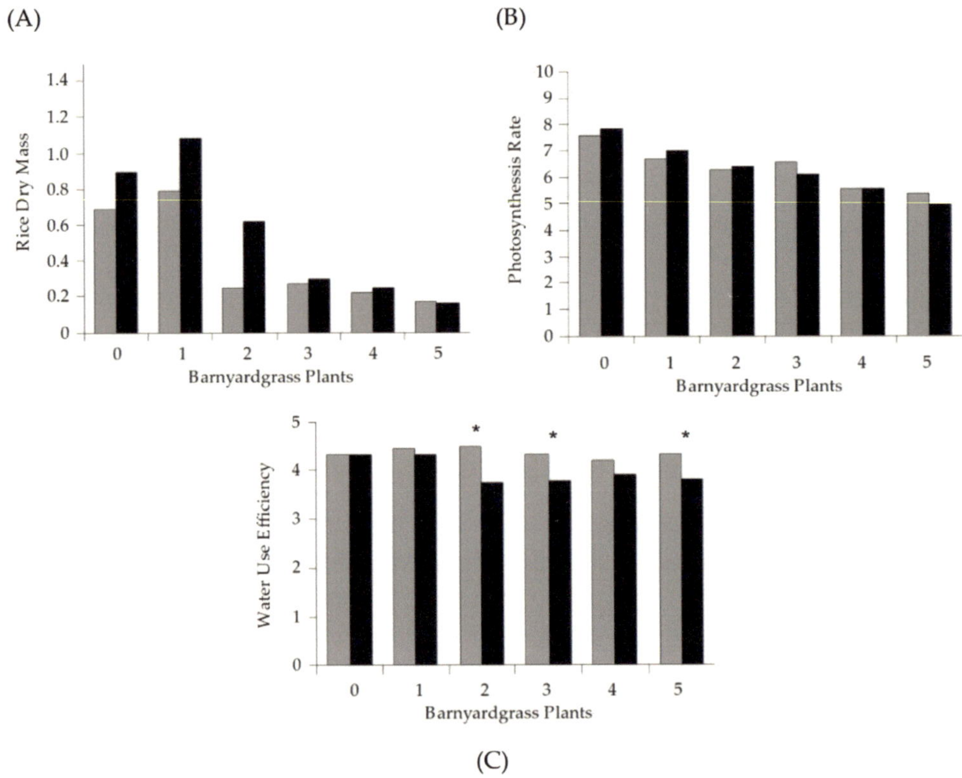

Figure 2. Dry mass (g $plant^{-1}$) (A), photosynthesis (μmol m^{-2} s^{-1}) (B), and water use efficiency (mol CO_2 mol H_2O^{-1}) (C) of rice plants variety BRS-Pelota as single plant, as function of competition with different numbers of barnyardgrass plants from the resistant (■) or susceptible (■) biotypes to Quinclorac. Source adapted from [37]. * Biotypes differ according to the LSD test at 5% probability.

Rice dry mass under competition with barnyardgrass did not differ from the plant free from competition, when competing with only one barnyardgrass plant. However, under competition with two or more barnyardgrass plants, the rice reduced its accumulation of dry mass per plant (**Figure 2**). Thus, the level of competition between the rice and the surrounding weeds is more important than the probable differences that may occur between biotypes or ecotypes of the same weed species [3].

In Brazil, Andres et al. were pioneers in collaborating with rice breeding programs in trying to identify those rice lineages from Embrapa Clima Temperado's rice breeding program that coupled superior ability to outstand emergence under unfavorable conditions (**Figure 3**) and superior ability to compete with weeds under field conditions (**Figure 4**) [35].

The superiority of genotype #19 is clearly visible in keeping seed emergence and vigor even after several days under moderately inadequate germination conditions (**Figure 4**). Furthermore, this genotype was also superior in a subsequent competition study by the substitutive method installed under field conditions, whose competitor species was exclusively the barnyardgrass (**Figure 5**).

Most academic studies that find significant differences on the impact of distinct biotypes of the same weed on rice are probably due to the application of inadequate experimental and statistical methods [39]. Most of these studies adopt designs that simply lack enough statistical power to identify any real difference between plant biotypes. Only few significant studies notify real differences on morphophysiological traits among biotypes.

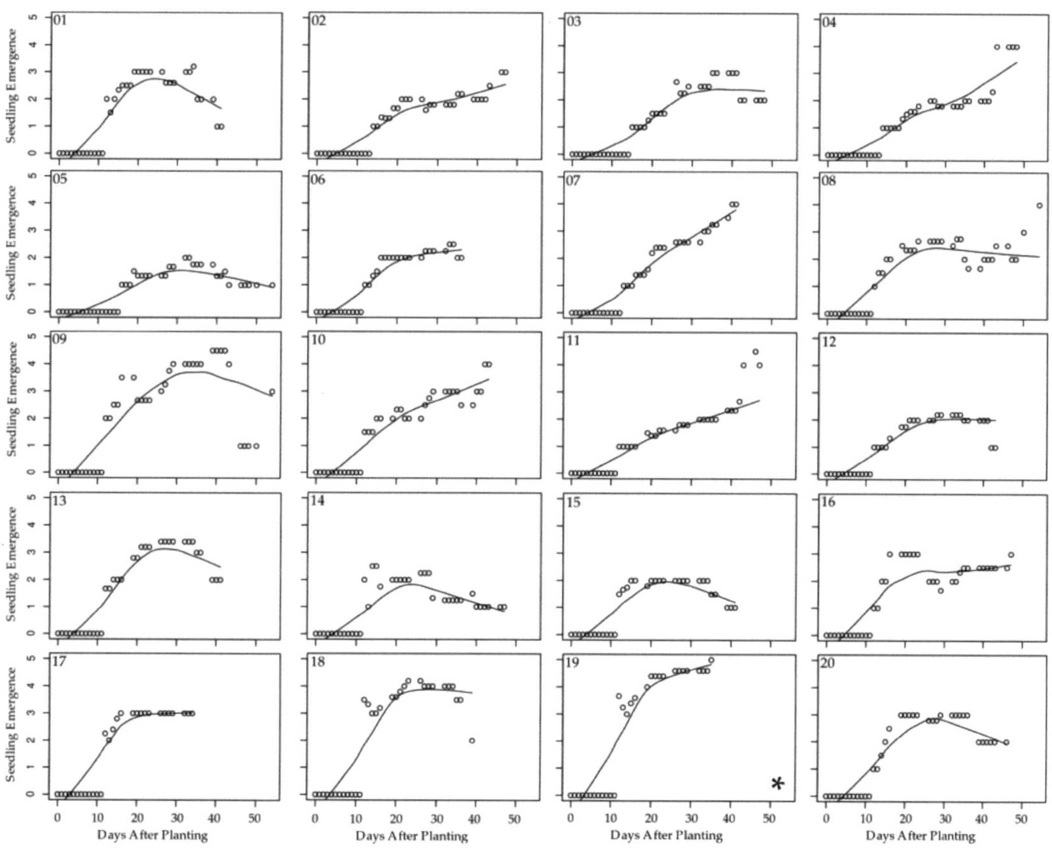

Figure 3. Emergence speed of 20 rice lineages randomly selected from the Embrapa Clima Temperado's irrigated rice breeding program. Source: adapted from [35].

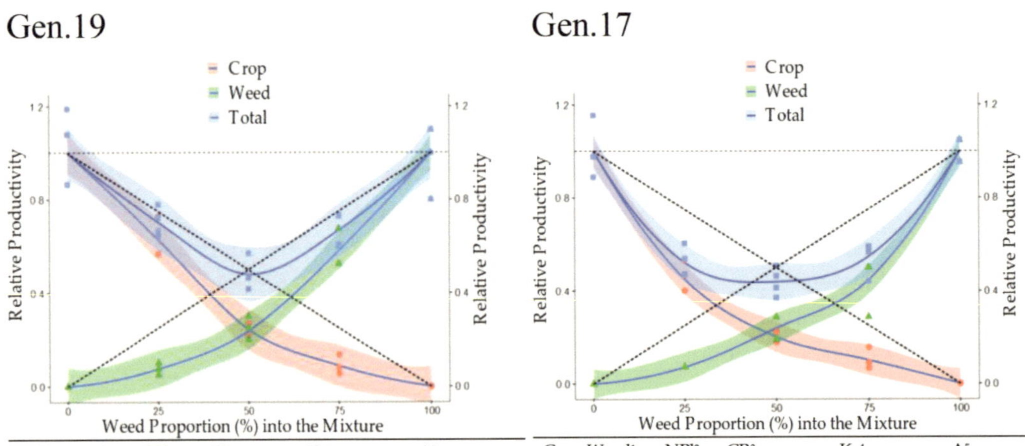

Crop:Weed[1]	NPI[2]	CR[3]		Kc[4]		A[5]	
100:0(F)	9.25						
75:25	7.75	3.041	*	0.598	*	0.269	*
50:50	4.5	1.05	ns	0.323	ns	0.006	ns
25:75	3.25	0.461	*	0.292	ns	-0.212	*
0:100	0						
C.V.	20.4	%					

Crop:Weed[1]	NPI[2]	CR[3]		Kc[4]		A[5]	
100:0(F)	11.25						
75:25	6.75	2.1	*	0.279	ns	0.157	*
50:50	4.5	0.875	ns	0.251	ns	-0.038	ns
25:75	4.5	0.775	ns	0.338	ns	-0.098	ns
0:100	0						
C.V.	20	%					

[1] plant proportion (crop:weed), being (F) the control plot free from competition; [2] crop emergence at the indicated competition level compared to the (F) plot by Dunnett's test; [3] significant when it differed from "1" according to the T-test; [4] difference between K_{crop} and K_{weed} (K_{weed} not shown) at the same competition levels, according to the T-test with Welch criteria; [5] significant when differed from "0" according to the T-test; * = significant difference at 5% level; [ns] = non-significant.

Figure 4. Relative performance of rice emergence under competition with barnyardgrass in field conditions, by the substitutive method of study. Source: adapted from [35].

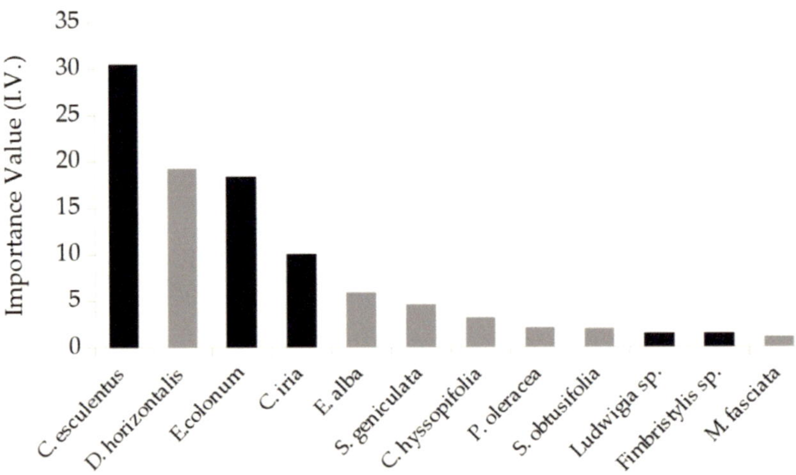

Figure 5. Importance value for weed species in areas with rice-soybean rotation for at least 5 years, in the Center-West region of Brazil. Source: adapted from [38].

Anyway, it seems not very wise to assign big efforts and resources in trying to differentiate weed biotypes in terms of their competitive ability with rice; there is a most urgent need in characterizing the main weed species traits, which confer them superior ability to compete with rice. Furthermore, it is also wise to focus on rice breeding programs that will select not only the most productive rice lineages for becoming commercial varieties, but that also associate to this important trait the presence of the morphophysiological features that are known to confer superior competitive ability to rice against the most important weed species occurring in this crop. This will confer stability to the rice grain yield and may help reducing the demand for herbicides as well [7, 8, 13].

Thus, the study illustrated in **Figure 2** shows that the proper moment for applying weed control techniques in rice is probably the same for any barnyardgrass biotype, and most probably the same tends to occur with biotypes of other weed species. The theory of weed impact on crops based on old ecological concepts like the *Critical Period for Weed Competition* (CPWC) [40], although applicable to the time they were adapted from ecological concepts, need to be at least partially reviewed considering the number of cases of weed biotypes with resistance to herbicides that appear every year [41].

In CPWC, there seems to be a most urgent need to revise the *Period Prior to Interference* (PPI) concept: considering the current case scenario regarding the difficulty to reach proper weed control with herbicides, it seems most prudent to always start the fields clean from any weed species; thus, the PPI would always be equal to zero.

4. Synecology of weeds in rice crops

The principles of the application of synecological methods to the weed science were summarized by Concenço et al. and will not be discussed in the present chapter [40]. We will focus on the application of synecology to rice fields.

Weed density or abundance is expressed by the number of plants within each quadrat. The density information collected in each quadrat is normally extrapolated to bigger areas. On the other hand, frequency is the proportion of total quadrats containing the individuals of the same species. If this species covers the most basal area of the community, it is considered a Dominant species.

According to several parameters (density, frequency, and dominance), the Importance Value (iv) of each species in the community can be easily estimated. Thus, the species that is present in higher density and frequency and is capable of suppressing other species due to a faster growth and mass accumulation (dominant) is the most important species (iv) within a plant community.

Weed composition of rice fields located in the lowlands of Southern Brazil was assessed and the authors reported the list of weed species found highlighting the ones with the highest potential to interfere in rice growth and development, thus impacting rice grain yield. The main results of this study are summarized in **Table 1** [39].

Barnyardgrass (*E. colona*) was responsible for 22.6% of the importance value of infestation into the studied areas. If this weed is completely eliminated from the area, the problems with weeds should be reduced with 22%, until another species takes advantage of physical

Weed species	Control	Clom. + Cyhal.	Clom. + Cyhal.	Penox. + Cyhal.	Penox. + Cyhal.	Imazet. + Imazap. + Cyhal.	Imazet. + Imazap. + Imazet. + Imazap.	Imazapyr + Imazap. + Cyhal.
Aeschynomene denticulata	0	0	0	0	5.73	0	0	0
Alternanthera filoxera	0	0	6.16	0	0	0	0	0
Brachiaria sp.	0	1.58	0	0	1.91	0	4.24	0
Conyza canadensis	0	0	0	5.03	3.9	0	3.67	0
Cynodon dactylon	18.92	10.48	2.68	10.71	11.3	13.27	26.92	21.57
Cyperus distans	0	0	10.79	0	0	0	0	0
Cyperus esculentus	12.83	29.38	18.99	41.5	25.92	30.26	20.52	22.27
Cyperus odoratus	3.47	2.62	0	0	0	0	0	0
Echinochloa crus-galli	22.57	15.76	5.19	4.92	5.35	8.57	0	8.55
Eleocharis elegans	0	0	0	0	0	4.14	0	0
Eleocharis sp.	4.56	1.88	0	7.06	1.68	2.71	0	0
Fimbristylis autumnalis	0	0	0	7.69	0	0	0	0
Fimbristylis sp.	14.64	9.3	20.27	9.44	18.94	31.1	18.46	11.96
Hypochaeris sp.	0	0	0	0	1.63	0	0	0
Kyllinga brevifolia	19.24	7.05	24.09	5.76	15.19	4.23	8.71	19.72
Lolium multiflorum	1.83	3.41	9.44	0	5.33	0	0	3.96
Oryza sativa	0	1.57	2.4	4.34	0	3	0	0

Weed species	Control	Clom. + Cyhal.	Clom. + Cyhal.	Penox. + Cyhal.	Penox. + Cyhal.	Imazet. + Imazap. + Cyhal.	Imazet. + Imazap. + Imazet. + Imazap.	Imazapyr + Imazap. + Cyhal.
Paspalum notatum	0	0	0	0	0	0	0	3.04
Pluchea sagittalis	1.94	0	0	0	1.64	2.73	11.44	5.63
Polygonum hydropiperoides	0	10.67	0	1.61	1.47	0	0	3.29
Porophyllum sp.	0	6.29	0	0	0	0	0	0
Rhynchospora sp.	0	0	0	0	0	0	3.31	0
Spermacoce capitata	0	0	0	0	0	0	0	3.29
Trifolium sp.	0	0	0	1.92	0	0	0	0

Source: adapted from Concenço et al [39].

Table 1. Importance value (iv—%) of weed species in rice fields in Southern Brazil, as a function of herbicide treatments.

Cultivar	Cycle	Clearfield®	Resistance to low temperatures	Resistance to diseases[1]	Resistance to insects[2]	Resistant to lodging	Susceptibility to iron toxicity	Grain quality	Productivity
EPAGRI 106	Early	NO	NI[3]	High	NI	Intermediate	Intermediate	NI	NI
EPAGRI 108	Late	NO	NI	Intermediate	NI	High	Low	High	High
EPAGRI 109	Late	NO	NI	Intermediate	NI	High	Low	High	High
SCS 114 Andosan	Late	NO	NO	Intermediate	NI	High	Intermediate	NI	NI
SCS 115 CL	Intermediate	YES	NI	Intermediate	NI	Low	High	High	NI
SCS 116 Satoru	Late	NO	YES	Intermediate	NI	NI	Intermediate	High	High
SCS117 CL	Late	YES	YES	Intermediate	NI	NI	High	High	High
BR-IRGA 409	Intermediate	NO	NI	NI	NI	NI	NI	High	High
BR-IRGA 410	Intermediate	NO	YES	Low	NI	NI	Low	Intermediate	High
BR-IRGA 414	Early	NO	NI	Low	NI	NI	High	High	High
BRS 6 Chuí	Early	NO	NI	NI	NI	NI	Intermediate	High	High
BRS 7 Taim	Intermediate	NO	NI	Intermediate	NI	NI	NI	High	High
BRS Atalanta	Very early	NO	NI	NI	Intermediate	NI	NI	High	High
BRS Querência	Early	NO	NI	Intermediate	NI	NI	NI	High	High
BRS Sinuelo CL	Intermediate	YES	NI	High	NI	High	NI	High	High
BRS Pampa	Early	NO	NI	High	NI	High	NI	High	High
IRGA 416	Early	NO	NI	High	NI	NI	NI	High	High

http://dx.doi.org/10.5772/intechopen.78342

Cultivar	Cycle	Clearfield®	Resistance to low temperatures	Resistance to diseases[1]	Resistance to insects[2]	Resistant to lodging	Susceptibility to iron toxicity	Grain quality	Productivity
IRGA 417	Early	NO	YES	High	NI	NI	High	High	High
IRGA 421	Very early	NO	NI	Low	NI	NI	Intermediate	High	High
IRGA 422 CL	Early	YES	NI	NI	NI	NI	NI	High	High
IRGA 423	Early	NO	NI	High	NI	NI	High	High	High
IRGA 424	Intermediate	NO	YES	High	NI	NI	High	Intermediate	High
IRGA 426	Intermediate	NO	NI	High	NI	High	Intermediate	High	High
IRGA 427	Intermediate	NO	NI	Intermediate	NI	High	High	High	High
IRGA 428	Intermediate	YES	NI	Intermediate	NI	NI	High	High	High
PUITÁ INTA-CL	Intermediate	YES	NI	NI	NI	NI	Intermediate	High	High
GURI INTA CL	Intermediate	YES	YES	NI	NI	NI	Intermediate	High	High
Avaxi CL	Early	YES	YES	High	NI	NI	High	High	High
Inov CL	Early	YES	YES	High	NI	NI	High	High	High

[1]Resistance to *Pyricularia grisea*.

[2]Resistance to *Oryzophagus oryzae*.

[3]NI—Information is not available.

Table 2. List of the currently used rice cultivars in Southern Brazil and their key physiological traits [31].

space made available by the control of the previous species. This weed was followed by *C. esculentus*, *C. dactylon*, and *Fimbristylis* sp. (**Table 1**) but the herbicides adopted in such rice fields are efficient against these species as well. It is also possible to observe that all herbicides were efficient in controlling barnyardgrass (treatments 2–8) (**Table 1**). Moreover, any other weed species was capable to take the place of barnyardgrass when the herbicides were applied. However, *C. esculentus* had increased in all treatments from 2 to 8, depending on the applied dose. This species was also dominant at T4, but with low frequency (**Table 1**).

In another study that was conducted in the lowlands of Center-West Brazilian region, Erasmo et al. found a similar composition of important weed species in rice fields installed in rotation with soybean for at least 5 years (**Figure 5**). In such study, *C. esculentus*, *E. colona* and *Fimbristylis* sp. were also among the most important weed species, similarly to what was observed in Southern lowland rice fields [39].

This raises a series of questions: (1) the most commonly herbicides used for weed control in rice may not be as effective on these weed species; (2) these weed species have similar demands for edaphoclimatic and nutritional resources to rice, thus adapting to the same environments; (3) it seems that soybean, when included into a crop rotation scheme with rice, may be not as effective in helping controlling rice weed species as anticipated by some authors on the long-run. These aspects need to be elucidated in future synecological studies.

Synecological studies are the first step for developing successful and competitive rice varieties against weeds since they allow for clearly identifying those species that are most harmful to rice. The further steps would include dedicated autoecological studies on these species and rice, and later breeding programs aiming to select the most significant features of rice varieties, which would make it most competitive against the weeds originally identified.

In Southern Brazil, there are several rice cultivars that have different physiological traits that can be already explored to increase crop competitive ability (**Table 2**). Even though these cultivars were not breed to compete with weeds, some of them have interesting features that allow crop plants to grow more rapidly and healthier, such as resistance to diseases, insects, iron toxicity, increasing their ability to compete with weeds in various environmental conditions. Moreover, the cultivars that perform better when weeds are present in the cropping fields with great biomass production and high yields should be selected for future breeding programs aiming to produce cultivars with high competitive ability against weeds.

5. Final considerations

In addition to the introduction of more competitive rice cultivars against weeds, the crop can also obtain some advantage when other cultural methods are manipulated, such as the adoption of higher sowing density [42, 43]. In a cropping field, the density of the weed community tends to be much higher compared to cultivated species. Thus, it is common to assume that weeds have higher competitive ability than the crop; however, this effect could be caused by the higher weed densities and not the real competitive potential of these species.

It is important to mention that moving forward for the understanding of crop competitive ability it is relevant to include other factors and variables in competition experiments, excluding the ones in which only the degree of competition varies and growth suppression are evaluated. It is also important to learn about individual morphophysiological traits, especially about root competition. Moreover, understanding the link between genetic traits and competitive ability is essential to ensure that a crop cultivar is competitive and productive.

Author details

Fábio Schreiber[1]*, Ananda Scherner[1], André Andres[2], Germani Concenço[2] and Francisco Goulart[1]

*Address all correspondence to: schreiberbr@gmail.com

1 Federal University of Pelotas, Brazil

2 Embrapa Temperate Agriculture, Pelotas, Brazil

References

[1] Balbinot AA Jr, Fleck NG, Agostinetto D, Rizzardi MA, Merotto A Jr, Vidal RA. Velocidade de emergência e crescimento inicial de cultivares de arroz irrigado influenciando a competitividade com as plantas daninhas. Planta Daninha. 2001;**19**:305-316

[2] Weiner J. Competition among plants. Treballs de la SCB. 1993;**44**:99-109

[3] Keddy PA. Competition. 2nd ed. Dordrecht: Kluwer; 2001

[4] Andres A, Machado SLO. Plantas daninhas em arroz irrigado. In: Gomes AS, Magalhães AM Jr, editors. Arroz irrigado no Sul do Brasil. Brasília, DF: Embrapa Informação Tecnológica; 2004

[5] Oerke EC, Dehne HW. Safeguarding production-losses in major crops and the role of crop protection. Crop Protection. 2004;**23**:275-285

[6] Jha P, Kumar V, Godara RK, Chauman BS. Weed management using crop competition in the United States: A review. Crop Protection. 2016;**95**:31-37

[7] Gibson KD, Fischer AJ, Foin TC, Hill JE. Crop traits related to weed suppression in water-seeded rice (*Oryza sativa* L.). Weed Science and Allen Press. 2003;**51**:87-93

[8] Jordan N. Prospects for weed control through crop interference. Ecology Applied. 1993;**3**:84-91

[9] Gurevitch J, Scheiner SM, Fox GA. Ecologia Vegetal. 2nd ed. Porto Alegre, Brazil: Artmed. ISBN 978-853 6319-18-6; 2009

[10] Radosevich SR, Holt JS, Ghersa CM. Ecology of Weeds and Invasive Plants: Relationship to Agriculture and Natural Resource Management. 3rd ed. Hoboken, USA: John Wiley & Sons. ISBN 978-047-1767-79-4; 2007

[11] Barbour MG et al. Terrestrial Plant Ecology. Menlo Park: Benjamin/Cummings; 1998

[12] Floss EL. Fisiologia das Plantas Cultivadas. 4th ed. Passo Fundo, Brazil: Universidade de Passo Fundo. ISBN 978-857-5156-41-4; 2008

[13] Silva AA, Ferreira FA, Ferreira LR, Santos JB. Biologia de plantas daninhas. In: Silva AA, Silva JF, editors. Tópicos em Manejo de Plantas Daninhas. Viçosa, Brazil: Universidade Federal de Viçosa. ISBN 978-857-2692-75-5; 2007. pp. 17-61

[14] Severino FJ. Supressão da Infestação de Plantas Daninhas pelo Sistema de Produção de Integração Lavoura-Pecuária [thesis (doctorate)]. Escola Superior de Agricultura "Luiz de Queiroz". Piracicaba: Universidade de São Paulo; 2005. 113p

[15] Ceccon G. Milho safrinha com solo protegido e retorno econômico em Mato Grosso do Sul. Revista Plantio Direto. 2007;**16**(97):17-20. ISSN 16778081 (Janeiro/Fevereiro 2007)

[16] Galon L, Agostinetto L, Moraes PVD, Dal Magro T, Panozzo LE, Brandolt RR, Santos LS. Níveis de dano econômico para decisão de controle de capim-arroz (*Echinochloa* spp.) em arroz irrigado (*Oryza sativa*). Planta Daninha. 2007;**25**(4):709-718, ISSN 0100-8358 (October/December 2007)

[17] Fleck NG, Agostinetto D, Galon L, Schaedler CE. Competitividade relativa entre cultivares de arroz irrigado e biótipo de arroz-vermelho. Planta Daninha. 2008;**26**(1):101-111, ISSN 0100-8358 (January/March 2008)

[18] Bianchi MA, Fleck NG, Lamego FP, Agostinetto D. Papéis do arranjo de plantas e do cultivar de soja no resultado da interferência com plantas competidoras. Planta Daninha. 2010;**28**:979-991. ISSN 0100-8358. Special Number

[19] Dingkuhn M, Johnson DE, Sow A, Audebert YA. Relationships between upland rice canopy characteristics and weed competitiveness. Field Crops Research. 1999;**61**:79-95

[20] Kawano K, Gonzalez H, Lucena M. Intraspecific competition, with weeds, and spacing response in rice. Crop Science. 1974;**14**(8):41-845

[21] Fischer AJ, Ramirez HV, Gibson KD, Pinhjeiro SB. Competitiveness of semidwarf upland rice cultivars against palisadegrass (*Brachiaria brizantha*) and signalgrass (*B. decumbens*). Agronomy Journal. 2001;**93**:967-973

[22] Ruchel Q, Kaspary TE, Lamego FP, Santi AL, Basso CJ. Habilidade Competitiva. Cultivar Grandes Culturas. 2011;**147**:22-24

[23] Andrew I, Storkey J, Sparkes D. A review of the potential for competitive cereal cultivars as a tool in integrated weed management. Weed Research. 2015;**55**:239-248

[24] Chauman BS, Thierfelder C, Mhlanga B. Weed management in maize using crop competition: A review. Crop Protection. 2016;**88**:28-36

[25] Gaudet ICL, Keddy PA. A comparative approach to predicting competitive ability from plant traits. Nature. 1988;**334**:242-243

[26] Reich PB. Root–shoot relations: Optimality in acclimation and adaptation or the "Emperor's New Clothes"? In: Waisel Y, Eshel A, Kafkafi U, editors. Plant Roots: The Hidden Half. 3rd ed. New York, NY: Marcel Dekker; 2002

[27] Robinson D. Scaling the depths: Below-ground allocation in plants, forests and biomes. Functional Ecology. 2004;**18**:290-295

[28] Rizzardi MA, Fleck NG, Vidal RA, Merotto A Jr, Agostinetto D. Competição por recursos do solo entre ervas daninhas e culturas. Ciência Rural. 2001;**31**:707-714

[29] Givnish TJ. On the adaptive significance of leaf height in forest herbs. The American Naturalist. 1982;**120**:353-381

[30] Andres A, Concenço G, Schreiber F, Teló GM, Fagundes PRR, Magalhães AM Jr, Moisinho IS, Martins MB, Coradini MC. Habilidade competitiva de genótipos de arroz irrigado com capim-arroz [Competitive ability of rice genotypes against barnyardgrass]. Pelotas:Embrapa Clima Temperado (Boletim de Pesquisa e Desenvolvimento). 2017;**281**. 54p

[31] Sociedade Sul-Brasileira de arroz irrigado – SOSBAI. Arroz irrigado: Recomendações técnicas da pesquisa para o Sul do Brasil. Santa Maria: Universidade Federal de Santa Maria; 2016

[32] Procópio SO, Santos JB, Silva AA, Donagemma GK, Mendonça ESV. Ponto de murcha permanente de soja, feijão e plantas daninhas. Planta Daninha. 2004;**22**(1):35-41. ISSN 0100-8358 (January/March 2004)

[33] Larcher W. Ecofisiologia Vegetal. São Paulo, Brazil: Rima. ISBN 858-655-2038; 2004

[34] Santos JB, Procópio SO, Silva AA, Costa LC. Captação e aproveitamento da radiação solar pelas culturas da soja e do feijão e por plantas daninhas. Bragantia. 2003;**62**(1):147-153. ISSN 0006-8705 (Janeiro/Abril 2003)

[35] Sinclair TR, Bingham GE, Lemon ER, Allen H Jr. Water use efficiency of field-grown maize during moisture stress. Plant Physiology. 1975;**56**:245-249

[36] Melo PTBS, Schuch LOB, Assis F, Concenço G. Comportamento de populações de arroz irrigado em função das proporções de plantas originadas de sementes de alta e baixa qualidade fisiológica. Revista Brasileira de Sementes. 2006;**12**(1):37-43. ISSN 0101-3122

[37] Noldin JA, Concenço G, Eberhardt DS, Galon L, Andres A. Resistência de Echinochloa sp. ao herbicida quinclorac [Resistance of Echinochloa sp. to the herbicide quinclorac]. In: Agostinetto D, Vargas A, editors. Resistência de plantas daninhas a herbicidas no Brasil [Weed resistance to herbicides in Brazil]. Pelotas: UFPel; 2014:106-129

[38] Erasmo EAL, Pinheiro LLA, Costa NV. Phyto-sociological survey of weed communities in flooded rice areas cultivated under different management systems. Planta Daninha. 2004;**22**(2):195-201

[39] Concenço G, Andres A, Teló GM, Martins MB, Moisinho IS. Phytosociological characterization of weeds as a function of residual herbicides applied to rice grown under sprinkler irrigation. Experimental Agriculture. 2017;**53**:1-12

[40] Nieto HJ, Brondo MA, Gonzales JT. Critical periods of the crop growth cycle for competition from weeds. PANS (C). 1968;**14**:159-166

[41] Heap I. International survey on herbicide resistant. Weeds. 2017. Available at: www.weedscience.org [Accessed: Feb 18, 2018]

[42] Juraimi AS et al. Weed management in direct seeded rice culture: A review. Australian Journal of Crop Science. 2013;**7**:989-1002

[43] Bajwa AA, Walsh M, Chaunan BS. Weed management using crop competition in Australia. Crop Protection. May 2017;**95**:8-13

Chapter 5

Enhancing Productivity in Rice-Based Cropping Systems

Uppu Sai Sravan and Koti Venkata Ramana Murthy

Additional information is available at the end of the chapter

http://dx.doi.org/10.5772/intechopen.76904

Abstract

In India, the rice-based cropping system is a major food production system with rice as the first food crop. The cereal-based cropping system is low-yielding and highly nutrient exhaustive resulting in the declining of soil fertility. Summer/pre *kharif* fallowing leaves on the land fallow for entire season and production of the cropping system is declined. Hence, crops that can improve the fertility status should be included in the cropping system. Development of short duration thermal insensitive rice varieties has encouraged multiple cropping involving a wide range of crops. Diversification of rice-based cropping systems with inclusion of pulses/legumes and oilseeds in summer fallows is one of the options for horizontal expansion, as they are known to improve soil organic matter through biological nitrogen fixation, root exudates, leaf shedding and higher below ground biomass. The strategy for higher yields in the cropping system should be formulated using the combined application of organics, inorganics and biofertilizers coupled with the inclusion of crops in summer fallows for sustainable yields and preservation of soil health.

Keywords: cropping system, legumes, productivity, soil fertility, sustainability

1. Introduction

On 16 December 2002, the United Nations General Assembly declared the International Year of Rice (IYR) as the year 2004. The main theme of IYR, called "Rice is Life", resulted from the fact that rice-based cropping systems are indispensable to people, directly or indirectly for food security, poverty control and world's peace. For approximately 70% of world's population, rice is the second most important food crop, being cultivated in more than 100 countries in 163 million ha with current rice production of 740.9 million tons compared to the global

IntechOpen

© 2018 The Author(s). Licensee IntechOpen. This chapter is distributed under the terms of the Creative Commons Attribution License (http://creativecommons.org/licenses/by/3.0), which permits unrestricted use, distribution, and reproduction in any medium, provided the original work is properly cited.

demand of 765 million tons by 2025. Rice is cultivated two or three times in a year in diverse environments and cropping systems starting with sole cropping systems in rainfed and irrigated conditions (temperate and tropical regions) to predominant mono cropping in irrigated regions (at tropics) [1]. Among the rice growing countries, India ranks second in production (157.2 million tons) next to China. In India, rice is cultivated on 44.14 million ha obtaining a production of 106.65 million tons and yield of 2416 kg ha^{-1} [2]. Cereals are the most widespread group of crops across the world occupying 20% of the global land or 61% of the total cultivated land. About 2/3 of the world's cropland area is predominantly occupied by wheat, maize, barley, rice and millets. Rice is the second most important crop at global level (around 11% of global cultivated area) and is the most important crop in South and Southeast Asia being also cultivated in the Amazon Basin, the southern United States, and southern Australia [3]. Rice-based cropping systems (RBCS) are the major contributing food production systems with rice as the first food crop, forming an integral part of this system. Rice-rice system is followed in irrigated cropping while rice-pulse system is adopted in rainfed lowlands leaving land fallow during pre *kharif*/summer season. The constraints in rice production are the declining rate of growth during yield formation, shortage of labor, depletion of natural resources and environmental pollution. Hence, improving the productivity of rice-based systems would eradicate the hunger and poverty, and facilitate economic development and food security.

Development of varieties with better yields and response to fertilizers, and the excessive use of chemical fertilizers have increased the yields of both *kharif* rice and *rabi* crops in rice-based cropping systems. Cereal-cereal cropping systems are more exhaustive and resulted in negative nitrogen balances in soil due to its extensive depletion of nutrients from soil resulting in a declining of the system productivity, soil fertility and health, compared to cereal-legume and cereal-oilseed systems [4, 5]. Continuous imbalanced use of chemical fertilizers has resulted in declined soil productivity. Due to the introduction of short and medium duration rice varieties, multiple cropping and the diversification of RBCS were possible with inclusion of pulses, oilseeds and vegetables in summer/pre *kharif* season. This has been found more beneficial, providing enhanced productivity of system and improved soil fertility status than cereal-cereal sequence [6–8]. After the harvest of winter season crops, a short time period (80–90 days) is available until the next rainy season crop having the possibility to include fast growing crops. Inclusions of short duration green manures and grain legumes/pulses in RBCS have been widely investigated and reported [9–11]. Combined application of inorganics and organics along with biofertilizers should be considered for the cropping system within a particular agro-climatic region. The integrated nutrient management system is ideal for RBCS as the rice is predominantly grown under submerged-anaerobic conditions, which offers a wider scope for harnessing various nutrient sources. When biofertilizers and organic manures are applied along with the inorganics, their efficiency is improved and nutrients can be mineralized faster and made available to the plants [12].

2. Different rice-based cropping systems

The three main characteristics of this type of cropping system are: (1) the biological characteristics of the crop and its response and influence to the physicochemical and ecological environment, (2) the crop sequences in the system and (3) the management techniques applied in the system including the varieties of crop species [13]. Rice is the major crop in India being cultivated under both rainfed and irrigated conditions. Traditionally, the rice varieties/cultivars are tall, having long duration, being low yielding with a grain to straw ratio of 0.4 and are not well responsive to the applied inputs [14]. Development of short duration photo-insensitive, dwarf and input responsive high yielding rice varieties with a grain to straw ratio of 0.55 has encouraged the multiple cropping involving a wide range of crops. The selection of crops in cropping systems was mainly dependent on agro-climatic and socio-economic conditions of the region with rice as a main crop. The prominent rice-based cropping systems in India are rice-rice, rice-wheat, rice-pulse and rice-potato (**Table 1**). In India, particularly in Indo-Gangetic plains, the rice-wheat zone is a predominant system occupying about 13.5 million ha area accounting for 23 and 40% of total rice and wheat area, respectively [15, 16]. The predominance of rice-wheat system in the whole Indo-Gangetic plains zone is particularly due to compatibility of the two crops mainly during sowing times.

Agro-climatic region	Rainfall (mm)	Soils	Prominent cropping system
Western Himalayan (Himachal Pradesh, Jammu & Kashmir, Uttarakhand)	1650–2000	Hill and Sub-montane	Rice-wheat, Rice-potato-potato
Eastern Himalayan (Assam, North East states, West Bengal)	1840–3528	Red sandy, Laterite, hill, Alluvial	Rice-fallow, Rice-rice, Rice-pulses/oilseeds
Lower Gangetic Plain (West Bengal)	1302–1607	Alluvial, Red and yellow	Rice-rice, Rice-wheat, Rice-potato-jute/vegetables
Middle Gangetic Plain (Bihar, eastern Uttar Pradesh)	1211–1470	Alluvial, Tarai and Calcareous	Rice-wheat Rice-maize, Rice-potato-sunflower
Upper Gangetic Plain (central and western Uttar Pradesh)	721–979	Alluvial, Tarai	Rice-wheat Sugarcane-ratoon-wheat
Trans Gangetic Plain	360–890	Alluvial and Calcareous	Rice-wheat
Eastern plateau and Hills (Chhattisgarh, Jharkhand, Madhya Pradesh, Maharashtra)	1296–1436	Red, Yellow Laterite	Rice-blackgram Rice-niger/linseed Rice-Vegetables
East coast Plain and Hills (Andhra Pradesh, Odisha, Tamil Nadu, Puducherry)	780–1287	Deltaic alluvium, Red, Laterite	Rice-groundnut-greengram, Rice-greengram/blackgram, Rice-rice
West coast Plains and Hills (Goa, Maharashtra, Karnataka, Kerala, Tamil Nadu)	2226–3640	Coastal alluvium, Red, Laterite	Rice-rice
Andaman and Nicobar Island	1600–3000	Red	Rice-fallow

Source: [17].

Table 1. Rice-based cropping systems in different agro-climatic regions of India.

2.1. Effect of cereal-based cropping system on soil properties

The cereal-cereal cropping system is the most predominant in India and the reports have mentioned unsustainability and declining factors for productivity i.e., higher fertilizer dose is needed to obtain the required current yield level [18]. Puddling, which is essential for rice cultivation impoverishes soil physical condition, increases bulk density and reduces the hydraulic conductivity. Furthermore, this practice is energy-consuming, deteriorates the soil health for growing the succeeding crops [19–21]. Repeated cultivation of rice leads to the formation of hard-pan below the plow layer, deteriorates the soil structure, inhibits the root elongation and delays the planting of a succeeding crop [22]. Continuous rice cultivation for longer periods with poor crop management practices has often resulted in loss of soil fertility and in turn leads to multiple nutrient deficiencies [23, 24]. Under puddled conditions, rice undergoes several changes i.e., aerobic to anaerobic environment, resulting in several physical and electrochemical transformations. Puddling operation is water and energy-consuming, breaks the capillary pores, destroys the soil aggregates, disperses the fine clay particles and soil strength is lowered in the puddle- layer. Imbalanced use of N-fertilizer in rice may increase the leaching of nitrates beyond the root zone leading to the ground water pollution in rural areas [25].

2.2. Strategies for enhancing productivity in rice-based cropping system

Some of the potential strategies for sustaining the productivity of rice systems are: (i) reduction of the rice monoculture and diversification of the cropping system with pulses/oilseeds and (ii) enhancing of the input use efficiency in existing double and triple rice-based cropping systems through improved technology and management practices. Diversification includes vegetables, grain legumes, oilseeds and green manures, which improves the productivity, reduces the pest incidence and enhances the soil fertility and its physical properties by providing a break in soil submergence [26]. In addition, the balanced fertilizer use, the combined use of organics, the mineral fertilizers and bio-fertilizers and the inclusion of summer/pre *kharif* crops are the possible optimal agro-techniques for sustainable yields, improved fertilizer use efficiency and restoration of soil fertility in cereal-based cropping systems [27, 28].

2.2.1. Chemical fertilizers

Application of higher quantities of fertilizers than recommended rates, more particularly N in Indo-Gangetic plains to rice-wheat cropping system (RWCS) has stagnated/declined the yield levels. Approximately 1/3 of farmers that cultivate rice-wheat apply 180 kg N ha^{-1} to both rice and wheat compared to the recommended dose of 120 kg N ha^{-1}. Such indiscriminate use of N fertilizers has decreased the yields due to low nitrogen use efficiency (21–31%) and some amount of N were lost through excessive N losses, nitrate leaching and groundwater pollution [25, 29, 30]. Hence, balanced fertilizer use i.e., application of fertilizers in right proportion, right time and appropriate method and in an integrated manner are the promising agro-techniques for a higher use efficiency of applied fertilizers sustaining the productivity of RWCS [31]. Application of nitrogen in excess or the lack fertilizer compared to the optimal amount significantly affect both rice yield and quality,-. Consequently, the balanced crop nutrition is of utmost importance [32].

http://dx.doi.org/10.5772/intechopen.76904

2.2.2. *Organic manures and green manures*

In RBCS, the usage of organic sources of nutrients viz. organic manures and green manures area are rapidly declining. Organic manures are traditional sources of nutrients, which help in maintaining the soil fertility. Among the organic manures, farmyard manure (FYM) is the principal source and is commonly available to the local farmers. They are relatively cheap soil amendments, rich in nitrogen, helping in sustaining the soil fertility and protection of the environment. Organic manures contain plant nutrients, though in small quantities in comparison to the chemical fertilizers. The presence of growth hormones and enzymes make them essential for improvement of soil fertility and productivity. In addition to this, the organic manures help in improving the use efficiency of inorganic fertilizers. The supply of essential micronutrients through organic manures has also improved plant metabolic activities especially in the early vigorous growth of plant. Findings of [33, 34] showed that the application of farmyard manure up to 10 t ha^{-1} has significantly increased the rice growth and yield-contributing traits as well as the grain yield.

Green manure crops can be grown in the rice-based cropping system as they reduce soil pH, improve the soil fertility, water holding capacity and partially diminish the need of nitrogen fertilizer for rice crop. The green manures increase the efficiency of applied mineral fertilizer, help in availability of other plant nutrients and improve the contents of soil organic matter [35]. In rice-based system, the winter crops are usually harvested in the last fortnight of April or early May and rice is transplanted during the last fortnight of July or early August. This fallow period of about 80–90 days is sufficient for the growth of short duration and fast growing green manure crops [36, 37]. The incorporation of animal manure or green manure adds N to rice soils and increase the organic matter in soil. The organic materials viz. green manure, compost or animal manure, have low C-N ratio, supply 20–30% to the current rice crop and 40–60% is stored in the soil [38].

Continuous application of organic material for long periods results in an increased output of decomposed organic matter annually [45]. Application of green manures Sesbania and Crotalaria at 10 t ha^{-1} to rice has significantly increased the grain yield of rice by 1.6 and 1.1 t ha^{-1}, respectively compared to no green manure application [46]. The soil organic carbon has been improved with the integrated application of NPK and FYM at all locations (**Table 2**). In rice-wheat systems, soil organic carbon was improved from 18 to 62% with organic sources compared to chemical fertilizers [47, 48]. Soil organic carbon and productivity were improved with the combinations of organic and inorganic fertilizers [49–51]. At lower fertility, the green manures showed the maximum response than at higher fertility levels. Groundnut (*Arachis hypogaea*) pod yields were at maximum with 30:26:33 kg NPK ha^{-1} fertility level plus gypsum combined with the application of green manure to rice [46].

2.2.3. *Crop residues*

As the cost of chemical fertilizers has increased dramatically in recent years, farmers find difficulties or cannot afford to purchase them. Hence, alternatives to chemical fertilizers such as crop residues might be better options to meet N-fertilizer requirements of successive crops in the cropping system. On an average, 25% of the total nitrogen, 50% of total phosphorus and

Location	Cropping system	Initial	After 20 years (g kg^{-1})			References
		(g kg^{-1})	Control*	NPK	NPK + FYM	
Bhubaneswar, India	Rice-rice	2.7	4.1	5.9	7.6	[39]
Faizabad, India	Rice-wheat	3.7	1.9	4	5	[39]
Karnal, India	Fallow-rice-wheat	2.3	3	3.2	3.5	[39]
Pantnagar, India	Rice-wheat	14.8	4.9	8.4	14.9	[40]
Pantnagar, India	Rice-wheat-cowpea	14.8	6	9	14.4	[40]
Bhairahawa, Nepal	Rice-wheat	10.3	7.3	8.8	—	[41]
Barrackpore, India	Rice-wheat-jute	7.1	4	4.3	4.5	[42]
Ludhiana, India	Rice-wheat	1.8	2	3.7	—	[43]

*Control: no fertilizer addition. Source: [44].

Table 2. Soil organic carbon after 20 years of alternative fertility treatments in rice-based cropping systems in India and Nepal.

75% of total potassium in the crop harvest are retained in the residues. An estimated 377 million tons of crop residues per year are available in India. With the incorporation of green manure or crop residues, the organic matter has been improved and soil physical conditions has been altered i.e., decrease in bulk density, increase in total pore space, water stable aggregates and hydraulic conductivity [22]. Dhaincha (*Sesbania aculeata*), Sunnhemp (*Crotalaria juncea* Linn.), blackgram (*Vigna mungo* [L.]), cowpea (*Vigna unguiculata* [L.]) and greengram (*Vigna radiata* [L.]) are some of the important legumes used as green manure plants and they are adaptable to different rice-based cropping system. These legumes have the ability to fix atmospheric nitrogen and sustain the productivity and profitability in rice-based cropping systems [52]. Incorporation of Sunnhemp crop residues produced highest seed and haulm yields of rice fallow blackgram (**Table 3**). Yields of rice fallow blackgram with greengram and blackgram residue incorporation were better than with bhendi (*Abelmoschus esculentus*), sesame (*Sesamum indicum* Linn.) and clusterbean (*Cyamopsis tetragonoloba*) residue incorporation [8].

Retention of crop residues is more beneficial than inorganic fertilizers as the residues supply better nutrients through decomposition helping in improving soil organic matter, availability of nutrients and achieving sustainability of the crop production systems. The impact of residue incorporation on succeeding crops depends on the produced quantity of residues and time and method of incorporation [53]. Residue retention in mungbean (*Vigna radiata* [L.])–wheat rotation has increased yields of both crops and nitrogen balances of the crop rotation. Mungbean and lentil (*Lens culinaris*) residues returned to soil have fixed about 112 and 68 kg N ha^{-1}, respectively which has resulted in positive N balances (64 and 27 kg N ha^{-1}, respectively) of the cropping system and hence the fertilizer N requirement could be reduced [54].

2.2.4. Legumes in cropping systems

Legumes in rotation with cereals not only fix atmospheric N through biological nitrogen fixation but also enrich soil fertility, nutrient recycling from deeper soil layers, minimize soil

Cropping system	Seed yield (kg ha^{-1})	Haulm yield (kg ha^{-1})
T_1: Fallow-rice-rice fallow blackgram	225	462
T_2: Sunnhemp-rice-rice fallow blackgram	382	743
T_3: Greengram-rice-rice fallow blackgram	342	667
T_4: Blackgram-rice-rice fallow blackgram	329	643
T_5: Sesame-rice-rice fallow blackgram	278	551
T_6: Clusterbean-rice-rice fallow blackgram	263	538
T_7: Bhendi-rice-rice fallow blackgram	291	569
SEm	11.92	24.18
CD (p = 0.05)	35	72

Source: [8].

Table 3. Seed and haulm yield (kg ha^{-1}) of *rabi* rice fallow blackgram as influenced by pre *kharif* crops in rice-based cropping system.

compaction, increase in organic matter, reduce pest and disease incidence, promote mycorrhizal colonization and sustain the productivity of cereal-based cropping systems [5, 55]. Intensification of rice-wheat system with short-duration and uniform maturing summer legumes (cowpea and mungbean) has enhanced the productivity and profitability to achieve nutritional security of the system [56]. The legumes/pulses contribute to the sustainability of cropping systems through (1) biological nitrogen fixation, which supplies nitrogen to the system (2) diversification of cropping system, which reduces the disease, pest and weed incidence and (3) provide food and feed that are rich in protein [13]. It is clear that the soil fertility and the physical properties have been enhanced with use of the legumes/green manure crops [57]. The excess application of N-fertilizer has resulted in environmental pollution as large amounts of N were lost as a consequence that fertilizer use efficiencies are very low [58] suggesting that legumes should be used as potential N source for future cropping systems [59].

Rice-legume crop sequences are considered most productive crop sequence in southern part of India as legumes can fix atmospheric nitrogen and scavenge mineral nitrogen. Mineral N may be lost through denitrification or leaching under flooded condition [60]. Grain legumes shed their leaves near maturity and the above ground biomass after harvesting (seeds along with residues and roots) contains nitrogen, improving the soil nitrogen balance and productivity [61, 62]. The legume residues contain about 20–80 kg N ha^{-1} (about 70% of it is derived from biological nitrogen fixation) depending upon the type of crop and the full N benefits will be realized if all the residues are incorporated after harvesting the seed yield [55, 61]. Legumes can be grown as green manure, as catch crop during summer season [15] and the experiments from various countries showed that legumes have improved the soil fertility and erosion control, the socioeconomic benefits and can be included in the rice-based cropping system [53]. Therefore, the succeeding crop yields in the cropping system are higher when legumes are included [63, 64]. The results from experiments revealed that in each year the yields of rice were significantly ($p < 0.05$) higher in legumes than in the fallow-based rice-wheat system (**Table 4**).

Treatment	2001/2002			2002/2003			2003/2004			Average (2001/2002–2003/2004)		
	0N	+N	Mean*	0N	+N	Mean	0N	+N	Mean	0N	+N	Mean
Fallow	3642	4107	3875c	3808	4368	4088c	3273	3819	3546b	3574	4098	3836b
Mungbean	4556	5154	4855a	4245	5401	4823ab	3776	4692	4234a	4192	5082	4637a
Cowpea	4559	5008	4784a	4520	5398	4959a	3909	4809	4359a	4329	5072	4701a
Soybean	3784	4496	4140b	4411	4953	4682b	4277	4460	4369a	4157	4636	4397a
Sesbania	4222	5207	4715a	4952	5621	5287a	3803	4349	4076a	4326	5059	4692a
Pigeonpea	4443	5029	4736a	4130	4642	4386bc	3426	4128	3777b	4000	4600	4300a
Guar	4360	4951	4656a	4025	4704	4365bc	3563	4278	3921ab	3983	4644	4314a
Mean	4224B	4850A		4299B	5012A		3718B	4362A		4080B	4742A	

*Means followed by different letter(s) within columns for legumes or within row for fertilizer treatments differ significantly ($p < 0.05$). Source: [65].

Table 4. Rice grain yield (kg ha^{-1}) as influenced by green manure legumes and fertilizer N in rice-wheat system.

2.2.5. Biofertilizers

Application of biofertilizers in rice fields is gaining attention in recent times. These are alternative sources of nitrogen to chemical fertilizers being eco-friendly, fuel independent and cost effective helping in a better crop nutrient management. The ecological and agricultural importance of these organisms depends upon the ability of certain species to carry out both photosynthetic nitrogen fixation and proliferation in diverse habitats. BGA and *Azospirillum* are capable of growing under rice canopies and have been identified as prospective biofertilizers for wetland rice cultivation. Indeed, biofertilizers bring directly or indirectly certain changes in the physical, chemical and biological properties of the soil in rice fields, which are of agronomic importance. Inoculation of mycorrhizal fungi to upland rice has improved the growth as well as the nutrient acquisition [66, 67]. Rice yields (both grain and straw) are enhanced with the use of effective suitable microorganisms [68]. In rice field, bio-nutrient application containing *Pseudomonas mycostraw*, cyanobacteria and *Azospirillum* has enhanced soil organic carbon by 14–18% [69]. Combined application of bio-inoculants and crop residues retention provides positive C and N balance in soil in rice-legume-rice cropping systems [70]. An integrated nutrient management strategy i.e., organic manures, crop rotations with legumes and application of chemical fertilizers in balanced proportions will improve the soil fertility and sustainability of rice-based cropping systems [44].

2.2.6. Summer/pre kharif crops in rice-based cropping systems

In cereal-based cropping systems, rice is grown during the rainy season (*kharif*) from June to October and *rabi* crops during the winter season from November to March/April leaving the land fallow in-between the harvest of both crops. These cereal-based cropping system yields have stagnated or declined resulting in a serious threat to the sustainability of the crop rotation [16]. The productions of rice and wheat have to be increased by about 1.1 and 1.7% per annum, respectively for the next four decades to ensure food security in South Asia [71]. Hence, to meet the increasing cereal demand, for sustaining the productivity and improving resource use efficiency there is a need of crop intensification with green manures, legumes/pulses, oil seeds and vegetables in summer season [72]. The crop intensification with legume crops in rice-based cropping system constitutes a viable alternative to traditional practices such as the cultivation of winter wheat or leaving the land fallow [73]. The most important characteristics of green manure legumes are to produce higher biomass with leafy growth, good nodulation activity and considerable amount of nitrogen in short period. These crops produce economic yield, improve the profitability, economic condition of small and marginal farmers and their biomass/residues can be utilized as animal feeds and/or their residues can be used as nitrogen source for the monsoon crop [8]. Research from [37] reported that Sunnhemp green manure has produced higher crop residues/biomass, while bhendi has recorded superior seed yield (**Table 5**). In addition, these leguminous crops/green manures conserve the soil fertility and fix atmospheric nitrogen [74]. Early rains in summer generally start from the beginning of May and a sufficient amount of rainfall occurs during May to August. There is a gap of short period (80–90 days) between harvest of *rabi* crop and planting of *kharif* rice, which is sufficient to take up a short duration crop preceding to rice [37]. The choice of crops and species in RBCS is limited. However, the feasibility of inclusion of these crops in the cropping system is possible, if they are short duration, uniformly maturing, high yielding and disease resistant and improve the long-term productivity and economic viability of the system. Grain legumes are preferred because of their food value and nitrogen fixing abilities that are superior to green manure crops [75]. However, the adoption of pulses/legumes/green manures must consider the growing season, the cost of production, the rainfall and irrigation facilities [27, 75]. Therefore, the selection of crop and variety is mainly specific to the location. Hence, sesbania and Sunnhemp as green manure crops, cowpea, greengram and blackgram as potential grain legumes; clusterbean, bhendi and potato (*Solanum tuberosum* Linn.) as suitable vegetables; groundnut and sesamum as potential oilseeds are suitable in RBCS. In spite of beneficial and positive effects of summer crops in rice-based cropping systems, the adoption by farmers is slow due to the lack of seed availability, escalating production cost and increasing labor wages. Furthermore, previous reports showed that green manure application was not profitable [75]. Diversification of the system with legumes/pulses may enhance profitability, reduce pests and diseases, minimize the risks from fluctuating weather by varying planting and harvesting times. The cropping system yields, profitability and production efficiency of rice-based cropping system were superior when bhendi/blackgram was included during pre *kharif* season rather than leaving land fallow [76]. In recent years, natural resources viz. land, water and energy are reduced and resource use efficiency is an important aspect for considering the suitability of a cropping system [77]. Hence, choice of crop to be grown needs to be optimally planned to harvest the synergism among them for higher productivity of the system and efficient utilization of resource base [78]. Hence, efforts are needed to promote intensification of rice-based cropping system in the country with legumes/pulses, oilseeds and vegetable crops to meet the demand for these crops and for sustaining the productivity [72].

Pre *kharif* crops	Seed yield (kg ha^{-1})	Greengram equivalent yield (kg ha^{-1})	Crop residues on fresh weight basis (t ha^{-1})
Sunnhemp	—	—	30.11
Greengram	217	217	10.48
Blackgram	385	341	5.94
Sesame	139	205	3.79
Clusterbean	93	54	1.16
Bhendi	1743	915	4.73
SEm	—	27.46	0.85
CD (p = 0.05)	—	85.00	2.55

Source: [37].

Table 5. Seed, greengram equivalent yield (kg ha^{-1}) and crop residues on fresh weight basis (t ha^{-1}) of pre-*kharif* crops in rice-based cropping system.

3. Conclusions

Cereal-based cropping system is the most promising system for about 70% of the global population. The yields have stagnated in recent years with the cereal-cereal system and the land is fallow in the summer/pre *kharif* season. Hence, crop diversification with pulses, oilseeds and vegetables in summer season shows a lot of promises in alleviating the poverty, employment generation, ensuring balanced food supply, and improving productivity and sustainability of the cropping systems. Legume/pulse crops in crop rotation with cereal-based system and their crop residue incorporation in soil sustain the C and N dynamics in soil. The adoption of green manures/pulses/leguminous crops in nutrient exhaustive rice-based cropping system saves the nitrogen fertilizer to successive crops, increases in grain yields and profitability, decreases the soil pH and improves the soil structure. Improved short duration, high yielding varieties and remunerative prices of pulses/oilseeds/vegetables will encourage their adoption in the cropping systems. The adoption of the summer crops in the system will require a lot of adaptive research depending on the soil and environmental conditions of a particular region of cropping. The reduction in the use of chemical fertilizers and balanced supply of nutrients in an integrated manner through inorganics, organics and biofertilizers will enhance the yield and soil fertility. Currently, nutrient supply is mainly focused on the major nutrients i.e., nitrogen, phosphorus and potassium, but RBCS requires also micronutrients since the multi-nutrient deficiencies have been observed. Consequently, they must be considered in the fertilizer schemes.

4. Future research needed

The adoption of any technology in modern agriculture can be acceptable and adoptable by farmers only if it is economically viable. Future research should focus on problems for non-adoption of these technologies by farmers and find out suitable ways for their adoption. Next, the adoption of summer/pre *kharif* crops instead of leaving land fallow, which could be made possible through best suited cultivars, seed availability, reducing production cost and optimizing production technologies is of high interest. Research must be initiated considering farmers' voluntary participation, identifying the farmer's problems and develop the location-specific technologies by finding ways for easy integration. The new varieties or species of green manure/grain legume crops should be developed that has multiple uses such as fodder, feed, fuel, and other commercial products. The interaction between mineral fertilizers, organics and nitrogen fixing organisms needs further study as a way of achieving better integration of the nutrition systems for different crops.

Author details

Uppu Sai Sravan[1]* and Koti Venkata Ramana Murthy[2]

*Address all correspondence to: saisravan050@gmail.com

1 Central Research Institute for Dryland Agriculture (ICAR), Hyderabad, Telangana, India

2 Agricultural Research Station, Ragolu, Andhra Pradesh, India

References

[1] Laborte AG, Gutierrez MA, Balanza JG, Saito K, Zwart SJ, Boschetti M, et al. RiceAtlas, a spatial database of global rice calendars and production. Scientific Data. 2017;**4**:170074. DOI: 10.1038/sdata.2017.74

[2] Anonymous. Agricultural Statistics at a Glance, Directorate of Economics and Statistics, Department of Agriculture and Cooperation, Ministry of Agriculture, Government of India; 2015. pp. 83-85

[3] Leff B, Ramankutty N, Foley JA. Geographic distribution of major crops across the world. Global Biogeochemical Cycles. 2004;**18**:GB1009. DOI: 10.1029/2003GB002108

[4] Kumar A, Yadav DS. Effect of long-term fertilizer on soil and yield under rice-wheat cropping system. Journal of the Indian Society of Soil Science. 1993;**41**:178-180

[5] Singh RK, Bohra JS, Nath T, Singh Y, Singh K. Integrated assessment of diversification of rice-wheat cropping system in Indo-Gangetic plain. Archives of Agronomy and Soil Science. 2011;**57**:489-506. DOI: 10.1080/03650341003641771

[6] Kumpawat BS. Production potential and economics of different crop sequences. Indian Journal of Agronomy. 2001;**46**:421-424

[7] Raskar BS, Bhoi PG. Producing and economics of winter sorghum (*Sorghum bicolor*)-summer vegetables cropping systems. Under irrigated conditions of western Maharashtra. Indian Journal of Agronomy. 2001;**46**:30-35

[8] Sravan US, Ramana Murthy KV, Siva GS. Sustainability of rice based cropping system through pre-*kharif* crops. Environment and Ecology. 2015;**33**:1121-1125

[9] Tiwari KN, Pathak AN, Tiwari SP. Fertilizer management in cropping system for increased efficiency. Fertilizer News. 1980;**22**:3-20

[10] Singh B, Singh Y, Khind CS, Meelu OP. Leaching losses of urea-N applied to permeable soils under lowland rice. Fertilizer Research. 1991;**28**:179-184

[11] Yadav RL, Dwivedi BS, Prasad K, Gangwar KS. Overview and prospects for enhancing residual benefits of legumes in rice and wheat cropping systems in India. In: Rao JDVK, Johansen C, Rego TJ, editors. Residual Effects of Legumes in Rice and Wheat Cropping Systems of the Indo-Gangetic Plain. New Delhi, India: Oxford/IBH Publishing Co; 1998. pp. 207-225

[12] Kundu DK, Pillai KG. Integrated nutrient supply system in rice and rice based cropping systems. Fertilizer News. 1992;**37**:35-41

[13] Jensen ES, Peoples MB, Nielsen HH. Faba bean in cropping systems. Field Crops Research. 2010;**115**:203-216

[14] Kataki PK, Hobbs P, Adhikary B. The Rice-wheat cropping system of South Asia. Journal of Crop Production. 2001;**3**:1-26. DOI: 10.1300/J144v03n02_01

[15] Singh VK, Sharma BB, Dwivedi BS. The impact of diversification of a rice-wheat cropping system on crop productivity and soil fertility. The Journal of Agricultural Science. 2002;**139**:405-412

[16] Ladha JK, Pathak H, Padre AT, Dave D, Gupta RK. Productivity trends in intensive rice-wheat cropping systems in Asia. In: Ladha JK et al., editors. Improving the Productivity and Sustainability of Rice-Wheat Systems: Issues and Impacts. ASA Spec. Publ. 65. ASA, CSSA and SSA. WI: Madison; 2003. pp. 45-76

[17] Gill MS, Shukla AK, Pandey PS. Yield, nutrient response and economic analysis of important cropping systems in India. Indian Journal of Fertilisers. 2008;**4**:11-48

[18] Yadav RL. Factor productivity trends in a rice-wheat cropping system under long-term use of chemical fertilizers. Experimental Agriculture. 1998;**34**:1-18

[19] Painuli DK, Woodhead T, Pagliai M. Effective use of energy and water in rice-soil puddling. Soil and Tillage Research. 1988;**12**:149-161

[20] Sharma PK, Ladha JK, Bhushan L. Soil physical effects of puddling in rice-wheat cropping systems. In: Ladha et al., editors. Improving the Productivity and Sustainability of Rice-Wheat Systems: Issues and Impacts. ASA Spec Publ. 65: ASA, CSSA and SSA. WI: Madison; 2003. pp. 97-113

[21] Tripathi RP, Gaur MK, Rawat MS. Puddling effect on soil physical properties and rice performance under shallow water table condition of tarai. Journal of the Indian Society of Soil Science. 2003;**51**:118-124

[22] Boparai BS, Singh Y, Sharma BD. Effect of green manuring with *Sesbania aculeate* on physical properties of soil and on growth of wheat in rice-wheat cropping systems in a semi-arid region of India. Arid Soil Research and Rehabilitation. 1992;**6**:135-143

[23] Fujisaka S, Harrington L, Hobbs P. Rice-wheat in South Asia: Systems and long-term priorities established through diagnostic research. Agricultural Systems. 1994;**46**:169-187

[24] Singh J, Singh JP. Land degradation and economic sustainability. Ecological Economics. 1995;**15**:77-86

[25] Singh B, Singh Y, Sekhon GS. Fertilizer N use efficiency and nitrate pollution of groundwater in developing countries. Journal of Contaminant Hydrology. 1995;**20**:167-184

[26] Cassman KG, Pingali PL. Intensification of irrigated rice systems: Learning from the past to meet future challenges. GeoJournal. 1995;**35**:299-305

[27] Timsina J, Connor DJ. Productivity and management of rice-wheat cropping systems: Issues and challenges. Field Crops Research. 2001;**69**:93-132

[28] Dwivedi BS, Shukla AK, Singh VK, Yadav RL. Improving nitrogen and phosphorus use efficiencies through inclusion of forage cowpea in the rice-wheat system in the Indo-Gangetic plains of India. Field Crops Research. 2003;**84**:399-418

[29] Singh Y, Khind CS, Singh B. Efficient management of leguminous green manures in wet land rice. Advances in Agronomy. 1991b;**45**:135-187

[30] Aulakh MS, Singh B. Nitrogen losses and fertilizer N use efficiency in irrigated porous soils. Nutrient Cycling in Agroecosystems. 1997;**47**:197-212

[31] Yadav RL, Prasad K, Gangwar KS, Dwivedi BS. Cropping systems and resource use efficiency. Indian Journal of Agricultural Sciences. 1998b;**68**:548-558

[32] Meena SL, Surendra S, Shivay YS, Singh S. Response of hybrid rice (*Oryza sativa*) to nitrogen and potassium application in sandy clay loam soils. Indian Journal of Agricultural Sciences. 2003;**73**:8-11

[33] Prakash YS, Bhadoria PBS, Rakshit A, Wais A. Response of basmati rice to integrated nutrient sources in lateritic soil of eastern India. Italian Journal of Agronomy. 2002;**6**:143-150

[34] Ghosh BN, Singh RD. Effect of conjoint use of farmyard manure and nitrogen on rice (*Oryza sativa*) - wheat (*Triticum aestivum*) system in Uttaranchal mid-hill soils. Indian Journal of Agricultural Sciences. 2003;**73**:680-683

[35] Saleem M, Akram M, Ihsan M, Akhtar, Ashraf M. Rice Production Hand Book. Islamabad: PARC; 2003. pp. 42-45

[36] Ali RI, Awan TH, Ahmad M, Saleem MU, Akhtar M. Diversification of rice-based cropping systems to improve soil fertility, sustainable productivity and economics. Journal of Animal and Plant Sciences. 2012;**22**:108-112

[37] Sravan US, Ramana Murthy KV. Scope of pre-*kharif* crops for productivity enhancement of rice based cropping system in north coastal zone of Andhra Pradesh. Research Journal of Agricultural Science. 2014;**5**:1100-1103

[38] Zhu Z, Liao X, Cal G, Chen R, Wang Z. On the improvement of the efficiency of nitrogen of chemical fertilizers and organic manures in rice production. Soil Science. 1983;**135**:35-39

[39] Lal R. The potential of carbon sequestration in soils of South Asia. In: Conserving Soil and Water for Society: Sharing Solutions. 13th International Soil Conservation Organisation Conference, Brisbane. 2004. pp. 1-6

[40] Ram N. Long-term effects of fertilizers on rice-wheat-cowpea productivity and soil properties in mollisols. In: Abrol IP, Bronson KF, Duxbury JM, Gupta RP, editors. Long-Term Soil Fertility Experiments in Rice-Wheat Cropping Systems. Rice-Wheat Consortium Paper Series 6. New Delhi, India: Rice-Wheat Consortium for the Indo-Gangetic Plains; 2000. pp. 50-55

[41] Regmi AP, Ladha JK, Pathak H, Pasuquin E, Bueno C, Dawe D, et al. Yield and soil fertility trends in a 20-year rice-rice-wheat experiment in Nepal. Soil Science Society of America Journal. 2002;**66**:857-867

[42] Saha MN, Saha AR, Mandal BC, Ray PK. Effects of long-term jute-rice-wheat cropping system on crop yields and soil fertility. In: Abrol IP, Bronson KF, Duxbury JM, Gupta RP, editors. Long-Term Soil Fertility Experiments in Rice-Wheat Cropping Systems. Rice-Wheat Consortium Paper Series 6. New Delhi, India: Rice-Wheat Consortium for the Indo-Gangetic Plains; 2000. pp. 94-104

[43] Rekhi RS, Benbi DK, Singh B. Effect of fertilizers and organic manures on crop yields and soil properties in rice-wheat cropping system. In: Abrol IP, Bronson KF, Duxbury JM, Gupta RP, editors. Long-Term Soil Fertility Experiments in Rice-Wheat Cropping Systems. Rice-Wheat Consortium Paper Series 6. New Delhi, India: Rice-Wheat Consortium for the Indo-Gangetic Plains; 2000. pp. 1-6

[44] Ghimire R, Lamichhane S, Acharya BS, Bista P, Sainju UM. Tillage, crop residue, and nutrient management effects on soil organic carbon sequestration in rice-based cropping systems: A review. Journal of Integrative Agriculture. 2016;**15**:60345-60347

[45] Suzuki M, Kamekawa M, Sekiya S, Shiga H. Effect of continuous application of organic or inorganic fertilizer for sixty years on soil fertility and rice yield in paddy field. In: Transactions of the 14th International Congress of Soil Science, Kyoto, Japan. 1990. pp. 1-19

[46] Prasad PVV, Satyanarayana V, Murthy VRK, Boote KJ. Maximizing yields in rice-groundnut cropping sequence through integrated nutrient management. Field Crops Research. 2002;**75**:9-21

[47] Gami SK, Ladha JK, Pathak H, Shah MP, Pasquin E, Pandey SP, et al. Long term changes in yield and soil fertility in a twenty year rice-wheat experiment in Nepal. Biology and Fertility of Soils. 2001;**34**:73-78

[48] Kukal SS, Rehana R, Benbi DK. Soil organic carbon sequestration in relation to organic and inorganic fertilization in rice-wheat and maize-wheat systems. Soil and Tillage Research. 2009;**102**:87-92

[49] Rasool R, Kukal SS, Hira GS. Soil physical fertility and crop performance as affected by long term application of FYM and inorganic fertilizers in rice-wheat system. Soil and Tillage Research. 2007;**96**:64-72

[50] Majumder B, Mandal B, Bandyopadhyay PK, Gangopadhyay A, Mani PK, Kundu AL, et al. Organic amendments influence soil organic carbon pools and rice-wheat productivity. Soil Science Society of America Journal. 2008;**72**:775-785

[51] Hossain MS, Hossain A, Sarkar MAR, Jahiruddin M, Teixeira da Silva JA, Hossain MI. Productivity and soil fertility of the rice-wheat system in the high Ganges River floodplain of Bangladesh is influenced by the inclusion of legumes and manure. Agriculture, Ecosystems and Environment. 2016;**218**:40-52

[52] Bar AR, Baggie I, Sanginga N. The use of sesbania (*Sesbania rostrata*) and urea in lowland rice production in Sierra Leone. Agroforestry Systems. 2000;**48**:111-118

[53] Singh Y, Singh B, Timsina J. Crop residue management for nutrient cycling and improving soil productivity in rice-based cropping systems in the tropics. Advances in Agronomy. 2005;**85**:269-407

[54] Shah Z, Shah SH, Peoples MB, Schwenke GD, Herridge DF. Crop residue and fertilizer N effects on nitrogen fixation and yields of legume-cereal rotations and soil organic fertility. Field Crops Research. 2003;**83**:1-11

[55] Wani SP, Rupela OP, Lee KK. Sustainable agriculture in the semi-arid tropics through biological nitrogen fixation in grain legumes. Plant and Soil. 1995;**174**:29-49

[56] Yaqub M, Mahmood T, Akhtar M, Iqbal MM, Ali S. Induction of mungbean (*Vigna radiata* (L.) wilczek) as a grain legume in the annual rice-wheat double cropping system. Pakistan Journal of Botany. 2010;**42**:3125-3135

[57] Ray SS, Gupta RP. Effect of green manuring and tillage practices on physical properties of puddled loam soil under rice-wheat cropping system. Journal of the Indian Society of Soil Science. 2001;**49**:670-678

[58] Crews TE, Peoples MB. Can the synchrony of nitrogen supply and crop demand be improved in legume and fertilizer-based agroecosystems? A review. Nutrient Cycling in Agroecosystems. 2005;**72**:101-120

[59] Crews TE, Peoples MB. Legume versus fertilizer sources of nitrogen: Ecological tradeoffs and human needs. Agriculture, Ecosystems & Environment. 2004;**102**:279-297

[60] Jeyabal A, Kuppuswamy G. Recycling of organic wastes for the production of vermicompost and its response in rice-legume cropping system and soil fertility. European Journal of Agronomy. 2001;**15**:153-170

[61] Giller KE. Nitrogen Fixation in Tropical Cropping Systems. Wallingford, UK: CAB International; 2001. p. 423

[62] Sharma AR, Behera UK. Recycling of legume residues for nitrogen economy and higher productivity in maize (*Zea mays*) -wheat (*Triticum aestivum*) cropping system. Nutrient Cycling in Agroecosystems. 2009;**83**:197-210

[63] Rochester IJ, Peoples MB, Hulugalle NR, Gault RR, Constable GA. Using legumes to enhance nitrogen fertility and improve soil condition in cotton cropping systems. Field Crops Research. 2001;**70**:27-41

[64] Ghosh PK, Bandypadhyay KK, Wanjari RH, Manna MC, Mishra AK, Mohanty M. Legume effect for enhancing productivity and nutrient use efficiency in major cropping systems-an Indian perspective: A review. Journal of Sustainable Agriculture. 2007;**30**:61-86. DOI: 10.1300/J064v30n01_07

[65] Shah Z, Ahmad SR, Rahman HU. Sustaining rice-wheat system through management of legumes I: Effect of green manure legumes on rice yield and soil quality. Pakistan Journal of Botany. 2011;**43**:1569-1574

[66] Gao X, Kuyper TW, Zou C, Zhang FS, Hoffland E. Mycorrhizal responsiveness of aerobic rice genotypes is negatively correlated with their zinc uptake when non-mycorrhizal. Plant and Soil. 2007;**290**:283-291. DOI: 10.1007/s11104-006-9160-x

[67] Zhang XH, Zhu YG, Chen BD, Lin AJ, Smith SE, Smith FA. Arbuscular mycorrhizal fungi contribute to resistance of upland rice to combined metal contamination of soil. Journal of Plant Nutrition. 2005;**28**:2065-2077

[68] Hussain T, Javaid T, Parr JF, Jilani G, Haq MA. Rice and wheat production in Pakistan with effective microorganisms. American Journal of Alternative Agriculture. 1999;**14**:30-36

[69] Jha M, Chourasia S, Sinha S. Microbial consortium for sustainable rice production. Agroecology and Sustainable Food Systems. 2013;**37**:340-362

[70] Thakuria D, Talukdar NC, Goswami C, Hazarika S, Kalita MC, Bending GD. Evaluation of rice-legume-rice cropping system on grain yield, nutrient uptake, nitrogen fixation and chemical, physical and biological properties of soil. Biology and Fertility of Soils. 2009;**45**:237-251

[71] Gathala MK, Kumar V, Sharma PC, Saharawat YS, Jat HS, Singh M, et al. Reprint of optimizing intensive cereal-based cropping systems addressing current and future drivers of agricultural change in the northwestern indo-Gangetic plains of India. Agriculture, Ecosystems and Environment. 2014;**187**:33-46

[72] Jat RA, Dungrani RA, Arvadia MK, Sahrawat KL. Diversification of rice (*Oryza sativa* L.)-based cropping systems for higher productivity, resource-use efficiency and economic returns in south Gujarat, India. Archives of Agronomy and Soil Science. 2012;**58**:561-572

[73] Schulz S, Keatinge JDH, Wells GJ. Productivity and residual effects of legumes in rice-based cropping systems in a warm-temperate environment I. Legume biomass production and N fixation. Field Crops Research. 1999;**61**:23-35

[74] Ladha JK, Kundu DK, Coppenolle MGAV, Peoples MB, Carangal VR, Dart PJ. Legume productivity and soil nitrogen dynamics in lowland rice-based cropping systems. Soil Science Society of America Journal. 1996;**60**:183-192

[75] Ali M. Evaluation of green manure technology in tropical lowland rice systems. Field Crops Research. 1999;**61**:61-78

[76] Sravan US, Ramana Murthy KV. Productivity and profitability of different rice (*Oryza sativa* L.) based cropping systems in north coastal zone of Andhra Pradesh. Ecology, Environment and Conservation. 2015;**21**:185-190

[77] Yadav JSP. Agricultural resource management in India: The challenges. Journal of Agricultural Water Management. 2002;**1**:61-69

[78] Anderson RI. Are some crops synergistic to following crops? Agronomy Journal. 2005; **97**:7-10

Chapter 6

Land Utilization Pattern in the Indonesian Forest: Cassava Cultivation in an Agroforestal System

Eva Banowati and Satya Budi Nugraha

Additional information is available at the end of the chapter

http://dx.doi.org/10.5772/intechopen.76928

Abstract

The potential forestland for agroforestry implementation in Indonesia is teak forest (*Tectona grandis*). The teak forest is less dense during the dry season, allowing sunlight to enter through the trees gap to the ground under the canopy. Therefore, some people use that condition as "palawija" farming land (palawija/phaladwija, in Java-Indonesia represents the type of non-rice agricultural crops). It is done to prevent the growth of weeds that can disturb the teak growth. The phenomenon of land utilization under the stands (PLDT) is an alternative in accessing forestland use by the community, a part of intercropping location. Theoretically, if the implementation was correct, it could be an effort to restore the forest ecological function. The pattern of the PLDT model on teak forests needs to select correct plants according to temporal dynamics, namely the season (dry or rainy) and the plants age. Land use representation could be seen from the cultivation pattern and crops variety that is cultivated under the forest stands at three research locations called Development Areas wilayah pengembangan (WP). The palawija crops that exist on all three WP were cassava (*Manihot esculenta* Crantz).

Keywords: agroforestry, teak forest, land utilization under the stands, intercropping, cassava

1. Introduction

Forestland intentional use for agriculture which combines some types of plants (trees and shrubs with crops or forage) is commonly called agroforestry [1–3]. In Indonesia, it is called "tumpangsari/wanatani," which is a type of land use pattern in forest, which combines forest and agriculture components (woody plants and annual crops) at the same time. The potential

IntechOpen

© 2018 The Author(s). Licensee IntechOpen. This chapter is distributed under the terms of the Creative Commons Attribution License (http://creativecommons.org/licenses/by/3.0), which permits unrestricted use, distribution, and reproduction in any medium, provided the original work is properly cited. (cc) BY forestland for agroforestry implementation is teak forest (*Tectona grandis*). It is one of the seasonal tropic forests, which is growing with the turn of the season (dry and rainy season). The teak forest is less dense during the dry season that allows sunlight to enter through the trees gap to the ground under the stands. Some people use that condition as "palawija" farming land (palawija/phaladwija, in Java-Indonesia represents the type of non-rice agricultural crops). It is done to prevent the growth of weeds that can disturb the teak growth.

The teak forests often form naturally due to the monsoon climate widely spread in the Northern Limestone Mountains, Kendeng Mountains and Muria Mountain. They also exist in Madura, Bali, Lampung (Sumatra), Flores (Nusa Tenggara Timur), Muna and some islands in Southeast Sulawesi. According to Banowati [4], the type of Javanese Teak forest spreads in Central Java and East Java Province. It can grow at an area up to an altitude of 200–650 m above sea level, with rainfalls of 1500–2000 mm per year and a temperature of 27–36°C, also dry months between 2 and 4 months. The best location for the teak growth is on soil with pH 4.5–7 that is not flooded with water (**Table 1**). The distribution of teak trees forms a tropical homogeneous forest in the limestone areas of Batang, Rembang, Blora, Grobogan, and Pati (Appendix 1).

The Indonesian teak forest is managed by the National Company of Forestry (Perum Perhutani) covering an area of 2.4 million hectares, consisting of protected forests (0.69 million hectares) and more than 1.72 million hectares (75.8%) as production forests. The extent of production forests allows the application of agroforestry patterns through Community Based Forest Resource Management pengelolaan hutan bersama masyarakat (PHBM) and the Land Use Model Under Stands pemanfaatan lahan di bawah tegakan (PLDT) program. The PLDT model is meaningful, especially in Java Island, because the people who live around the forests need farming land, which become narrower triggered by high population growth. Since 2001, the National Company of Forestry implemented the Community Based Forest Resource Management (PHBM). The PHBM paradigm was updated because it originally only prioritized the wood production, while using the word "forest," and changed to "forest resources."

Research location (village/plot)	Soil: solvent used			pH	Temp. (%)	Elevation (m)	Slope (%)
	Aα Bifiridil	H_2O_2	HCl				
I. Gesengan	No color changed	No froth	Many froths				
a. Plot 100				6.8	32°C/66	67	5
b. Plot 102				6.5	30°C/67	62	5
c. Plot 103				6.5	30°C/67	62	5
II. Semirejo	No color changed	No froth	Many froths				
a. Plot 114				6.8	29°C/72	138	16
b. Plot 115				6.6	30°C/67	137	10
III. Regaloh	No color changed	No froth	Many froths				
a. Plot 130				6.7	30°C/67	135	5
b. Plot 131				6.7	30°C/67	135	5

Source: [4].

Table 1. Physical measurements at the research locations in October 2016.

Through this program, the forest surrounding communities have the right to work the land with each covering area of 0.25–0.5 hectares as intercropping area to support the workers' economic activities and to maintain the ecological sustainability of the forest. They were coordinated in the institution of village forest community lembaga masyarakat desa hutan (LMDH).

Forestland has relatively more fertile soil because it can naturally conserve the soil fertility through the closed system of nutrient cycle. The nutrients used for growth can restore fertility into the soil through the fallen leaves, twigs and branches [5]. Forests are land cover that refers as a place of vegetation, which is influenced by soil type, as in the research location located in Muria Forest Area kawasan hutan muria (KHM) and spreaded on Volkan Muria landform. It consists of Red Yellow, Mediterranean and Latosol [6]. According to Nursanti [7], Latosol is a soil which has eroded intensively, acid reacted and washed strongly, especially for K, Ca, and Mg bases. Latosol soil type has a medium fertility level, and for agricultural cultivation, it lacks P nutrient due to fixation by kaolinite clay minerals and Fe ions. Moreover, it lacks Al ions due to the low pH level. However, with intensive soil management, Latosol soil fertility can be improved by planting long-term vegetation (forests) so that the nutrient availability could be increased.

The phenomenon of land utilization under stands (PLDT) is an alternative in accessing forestland use by the community, a part of intercropping location. In the beginning, the PLDT implementation by the forest surrounding community was done without a legal procedure. Theoretically, if the implementation was correct, it could be an effort to restore the forest ecological function. The correct concept of ecosystem in a tropical land cover area is the leaf type of grass, shrub and tree canopy (forest stands). The PLDT activity could control the growth potential of reed, which could be harmful for the soil. Without owning land, there is no certainty for the villagers (farmers) to meet the needs of their family members. Palawija could act as forest floor plant, which potentially helps in hampering the erosion rate improving the quality of land by plants that have root nodules that fix the nitrogen (N). The types and varieties of palawija crops were adjusted to the standing of biophysical condition, which refers to the concept of crop rotation, in order to form natural formations both vertically and horizontally. The plants formation in teak forests requires ecological engineering in line with silviculture principles in conserving forest resources to treat forests properly and control their structure and growth without jeopardizing their production capacity [4, 6–11].

The pattern of the PLDT model on teak forests needs to select correct plants according to the temporal dynamics, namely the season (dry or rainy) and the plants' age. Selection of cropping patterns and types of food crops needs to consider the difference of plant canopy density as described in [11] classified as dense enough (40–70%) and rare (less than 40%). These were the best spots to efficiently utilize and fulfill the requirements for sunlight, water and mineral nutrients. Furthermore, based on [12], the amount of sunlight that escapes through the canopy between November and December was 4.47–14.85% of the open light. It could reach the forest floor and could be harmful for the teak stands growth. In addition, it is necessary to note that the physical condition of the land must be measured and analyzed to determine the accurate planting patterns and select the correct plants.

Land use representation could be seen from the cultivation pattern and crops varieties that were cultivated under the forest stands at three research locations called Development Areas

(WP). The palawija crops that exist on all three WPs was cassava (*Manihot esculenta*) with different proportions.

2. The land use pattern under teak stands

The land use pattern under the stands at the first WP was monoculture planting pattern used in the intercropping system with cassava (*Manihot esculenta*) crops as the main crop. However, the pattern at the second and third WP was polyculture planting pattern used as compound system with cassava crops that were less than peanuts (*Arachis hypogaea*). Cassava planting on Latosol soils was not recommended because it was susceptible to P nutrient deficiency due to fixation by kaolinite clay minerals and Fe and Al ions due to low pH level. If the cassava cultivation was continuously done without proper soil maintenance or crops rotation, it could decrease the soil condition and the land potential.

The cassava crop as monoculture pattern at the research location was well grown. Farmers chose UJ-5 varieties of cassava (*Cassesart*) and *Margona* in other locations. They utilized post-harvest teak forestland as well as between the young teak stands. In **Figure 1**, the cassava crops have dense fresh green leaves, which indicated the young age (±4 months).

Land use in the second WP was polyculture cultivation pattern used as mix cropping system. It is a plant cultivation system, which mixes more than one crop at the same land and time. The land distribution for cropping system was unregulated without regarding the space between. This simplicity was based on farmers/community understanding that the forestland is fertile, and whatever was planted would be fruitful. Crops varieties including peanut and corn (*Zea mays*) are used as daily food needs, and the yield surplus is sold at the local market (**Figure 2**).

Farmers have learned to manage the land from the nature, which provided technical mind of local knowledge. It could be seen from the terrace that was built to manage the water supply and to overcome erosion rate that is caused by the agriculture land use (seasonal). It was built across the contour in an angle of 135°.

Figure 1. Monoculture of cassava crops at the post-harvest teak forest.

Figure 2. Mixed cropping system adapted to the land physical condition.

At the third WP, farmers have utilized farmland for agriculture using polyculture compound systems, and the main crop was the peanut. It was chosen related to its roots' ability to produce N element inside the soil that could be useful for Murbei (*Morus*) plants which has been cut down. The N element could help the Murbei growth immediately and produce quality and leafy leaves. Corn plants cropping as boundary plants showed more natural characteristics compared to non-plant boundaries. This local knowledge at the third WP can be seen in **Figure 3**.

Based on **Figures 1–3**, they indicate that the land use at the first WP is more oriented to economic aspects. The land use results meet the needs of tapioca flour industry. Different results at the second WP reflect more prioritized ecological aspects, while the farmland utilization at the third WP was a combination of economic and ecological aspects. Cassava was chosen as seasonal plants for agroforestry because it has the ability to stand against pests, have a simple vegetative breeding,

Figure 3. Compound system dominated by peanut.

and relatively stable price and the existence of tapioca flour industry as one of permanent stakeholder, which receives the cassava yield. The cassava cultivation was done throughout the year. The *Margona* variety has a planting age of 8–9 months, while the *Cassesart* was planted since 2014.

3. Cassava planting under the teak stands

Agricultural plants varieties which were cultivated at PLDT plots were cassava (*Manihot esculenta*), peanut (*Arachis hypogaea*), corn (*Zea mays*) and long beans (*Vigna unguiculata* ssp. *sesquipedalis*). Among the four crops, the cassava was cultivated in all WPs although the proportions were different, the other crops were only cultivated at the second and third WP. Some considerations of these diversities selection were influenced by the habit and general knowledge of the community. They were considered whether the plant could damage the environment or vice versa. They considered that cultivated land is better than empty land. On a wide scale, the third WP condition was in line with [13] which was done at a mountain village in Aga Khan, Pakistan. There was a tendency of harmonious relationship between the sustainability of biodiversity in a village and the community. There was no barrier for accessing the biodiversity. Development was succeeding to synergize economic and ecological functions. This condition could be followed by the first WP and become a consideration matter for the Government in determining the public policy related to the control of the state forestland, especially when distributed at the mountainous area.

Farmers cultivate cassava continuously and routinely, but the result was decreasing slowly because the farmlands was lacking of nutrients, which affected their productivity and cassava's quality, determining a low price, as well. The suggestion to not cultivate cassava in forestland was neglected by farmers for several reasons, that is, (1) financial limit to buy seeds, while for the cassava, it only needs the stem cutting, (2) the farmers are not brave enough to speculate on other plants; (3) worries about crops failure, while it needs high cost of maintenance, fertilizers and pesticides, (4) ease of cassava marketing, which is supported by the tapioca flour industry existence, and (5) simple method of cropping system.

Cassava planting was done by placing the stem at a depth of approximately 5 cm. The spacing of 9 × 9 cm produces 11,200 cassava trees per hectare. Weeds cleaning was done by cleaning the grasses without using pesticides and only once in each cropping cycle. Fertilization activities were done twice in each cropping cycle, at the beginning of the planting and at the third or fourth month after. Replanting process in this study was done when the cassava plants had too many buds, and if it was still in a reasonable condition, the farmer did not do it. The last activity was harvesting by removing the cassava trees at 9–12 months' age, depending on the seed used.

Based on the interpretation of SPOT 6 satellite image acquired in 2014, it was known that the cassava farmland distribution was on the north part of Muria Volcano, which 83.7% laid on the high slopes. It covered the Tlogowungu, Margoyoso, Cluwak, Gembong, Margorejo and Tayu districts. Meanwhile, 16.3% was on the south, which is the north part of Kendeng Mountain, including Sukolilo, Kayen and Tambakromo districts (**Figure 4**).

The actual area of cassava in 2014 was 18,544 ha, but the productive one was only 96.37% (17,871 ha) (**Table 2**). The total production of wet cassava and its cover was 744,746 tons [15].

http://dx.doi.org/10.5772/intechopen.76928

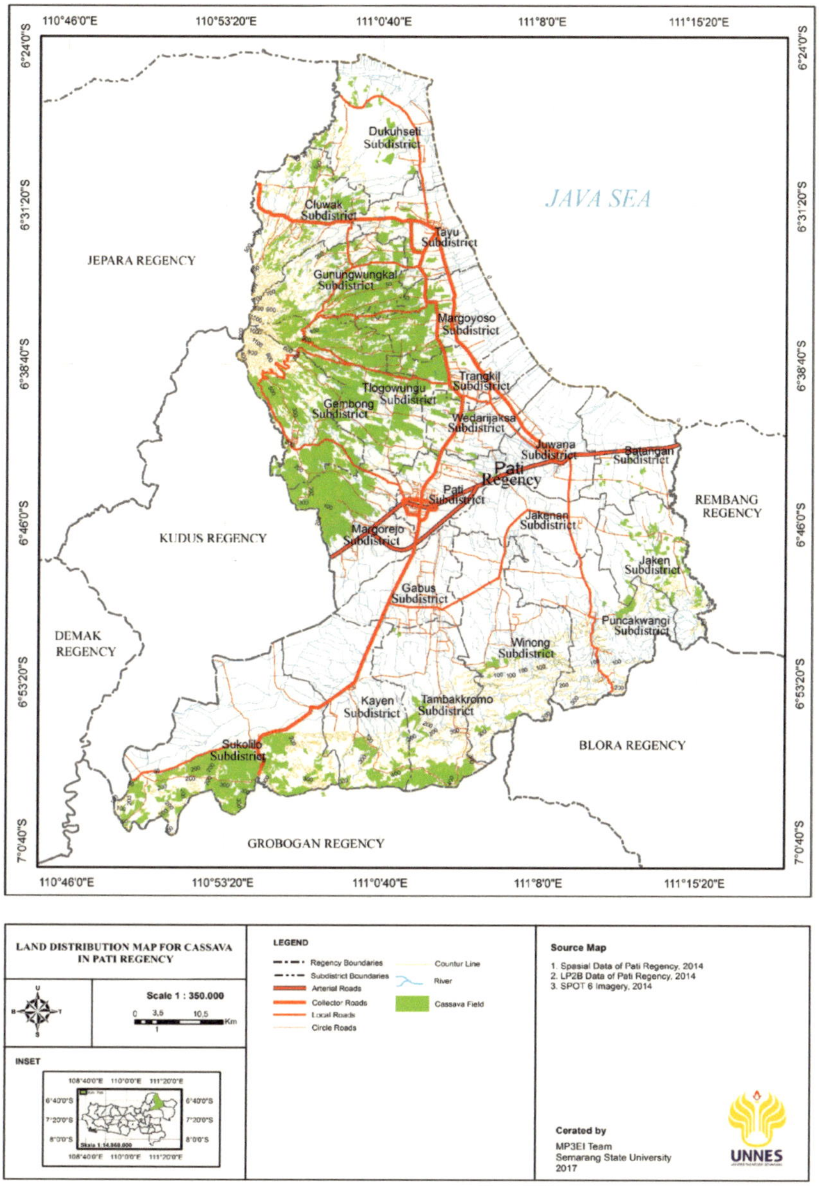

Figure 4. Map of cassava farming distribution in Pati Regency/east slope of Muria Forest. (Source: Spatial data of Pati Regency, 2014; LP2B data of Pati Regency, 2014; SPOT 6 imagery, 2014).

Nevertheless, tapioca processing industry did not run optimally due to instability of raw material supply of cassava because of several causes. On the one hand, one cause was farmer's reluctance because of the cassava price declining in the last 2 years (2014 and 2015). On the other hand, another cause was the farmer's side problem, including the reducing of farmer's land tenure, limited access to capital/financial support, non-innovative technologies to manage land and to process yields. Therefore, it is required to optimize farming productivity in terms of supply, demand and price fluctuations.

No	Year	Harvest area (ha)	Total production (ton)	Average production (ton ha^{-1})
1	2015	15,200	661,976	43.55
2	2014	17,871	744,746	41.67
3	2013	16,163	695,460	43.03
4	2012	19,696	732,961	37.21
5	2011	17,431	532,874	30.37
6	2010	21,989	643,558	29.27
7	2009	16,994	386,434	22.70
8	2008	16,740	318,194	19.00

Source: [14].

Table 2. Cassava production in Muria forest of Pati Regency.

Based on analysis result of spatial distribution pattern in **Figure 4**, it was shown that regional distribution index (Moran's Index) has influenced the production and supply continuity, also the transportation factors in terms of cost, distance and travel time (**Table 3**).

L value comes from the total number of joint areas; for example, in Tlogowungu district, the L value was 6, which means that the district is bordering to another six districts. The x value was 4283, which means that the total area of cassava farming land in that area was 4283 ha. While the L value of Pati District was 6), it means that transportation of yields was easy but the total area of cassava farming land was only 16 ha. Based on calculation result, it was found that the total joint area was 63 (**Table 3**). Calculation result determined the index of spatial distribution pattern using formula (1):

$$I = \frac{n \sum_{(c)} (x_i - \bar{x})(x_j - \bar{x})}{J \sum_{(c)} (x - \bar{x})} \tag{1}$$

$$I = \frac{18 \times 369407.9544}{63 \times 27753220}$$

$$I = \frac{6649343.179}{1748452.860} \quad I = 0.0038$$

Based on the formula, a positive I value of 0.0038 was obtained, which means that the cassava farming land distribution pattern in Pati Regency was clustered. This condition could facilitate the cassava yield transport to the local market and to collectors or brokers who supply the tapioca industry.

Referring to serial data of BPS Pati Regency, the cassava production in 2012 increased with 27.3% of total harvest area of 19,696 ha. However, it showed a declining trend pattern due to the decreasing cultivation area of about 9.27% (in 2014), and by 2015, the farming land decreased with 14.95% (15,200 ha of cropland). This affected the cassava production to 661,976

No	District	L	L^2	Region Value		
				(x)	$(x - \bar{x})$	$(x - \bar{x})^2$
1	Tlogowungu	6	36	4.283	3.417	11675.889
2	Gembong	2	4	3.276	2.410	5808.100
3	Cluwak	3	9	2.427	1.561	2436.721
4	Gunungwungkal	5	25	1.400	534	285.156
5	Margoyoso	4	16	1.097	231	53.361
6	Margorejo	4	16	1.638	772	595.984
7	Trangkil	3	9	537	−329	108.241
8	Dukuhseti	2	4	110	−756	571.536
9	Tayu	4	16	301	−565	319.225
10	Sukolilo	1	1	115	−751	564.001
11	Jaken	3	9	72	−794	630.436
12	Winong	4	16	51	−815	664.225
13	Tambakromo	3	9	51	−815	664.225
14	Kayen	3	9	63	−803	644.809
15	Wedarijaksa	4	16	85	−781	609.961
16	Pati	6	36	16	−850	722.500
17	Pucakwangi	3	9	53	−813	660.969
18	Batangan	3	9	7	−859	737.881
Total area join		63	249	15.582		27753.220
Rata-rata				866		

Source: ([14]); Secondary Data Analysis, 2017.
Information:
L = number of joint area.
x = total area of cassava farming land in each district.

Table 3. Spatial distribution pattern of cassava farming land (joint area analysis).

tons or 231691.6 tons of wet tapioca. Lack of cassava yields was overcome by supplying from other regions yields. At national scale, the total imports from January to April 2017 reached 1234 tons. Import of cassava was done to meet the need for less supply positions, which was triggered by the relatively low price of cassava. Therefore, farmers might switch to cultivate other crops (**Figure 5**).

Cassava supply from farmers tends to decline since 2014 as a result of farmer reaction who did not want to cultivate their land and delay the harvest time. This made the cassava market price to decline up to 44.6%. Tapioca industries did not produce optimally, but tapioca imports conducted by the Ministry of Trade, up to June were more than 1 million tons/year [14]. Domestic industries that use tapioca flour prefer imported tapioca flour due to cheaper price, better quality and continuous supply assurance [16]. However, the Ministry of Agriculture had cassava

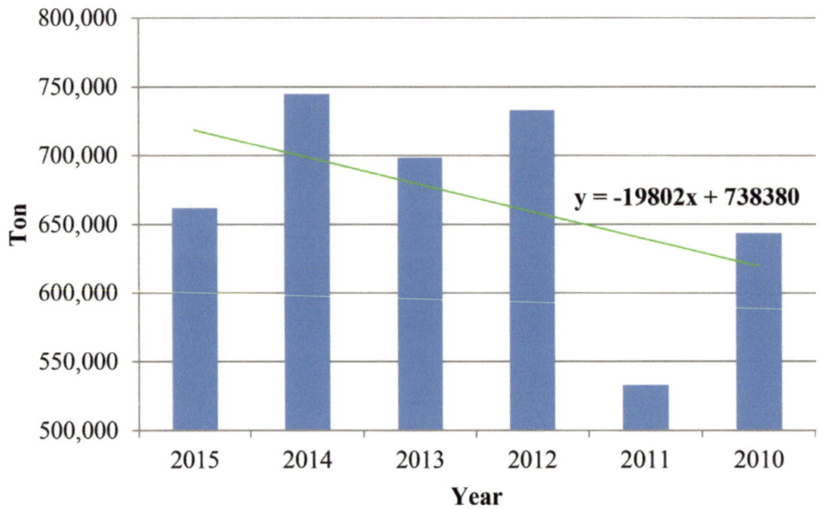

Figure 5. Trend of cassava production in Pati Regency. (Source: [14]; Secondary Data Analysis, 2017).

production data of local farmers, which are more than enough to meet local market needs. At national scale, other causes of imports are not influenced by lack of cassava production, but according to [16–19], not all cassava production meets the proper quality standard of Hazard Analysis Critical Control Point Specification (HACCP).

The highest production of cassava at national scale was in 2012, then decreased in years after. Different phenomenon occurred as compared to the production stability in Pati Regency, although the production decreased in 2015 as the effect of cassava price declining in the market. The cassava of Pati Regency contribution to food availability could be categorized in two levels, namely national and province scale (**Figure 6**).

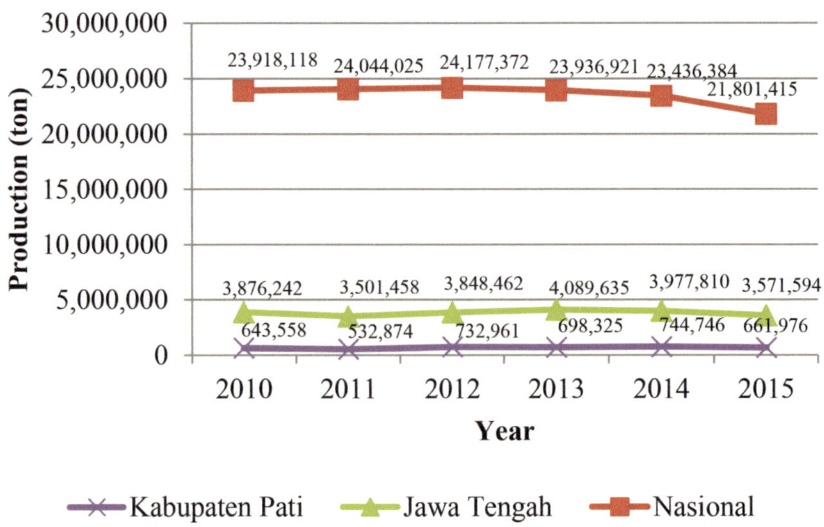

Figure 6. National cassava production in 2010–2015 (Source: [14]; Secondary Data Analysis, 2017).

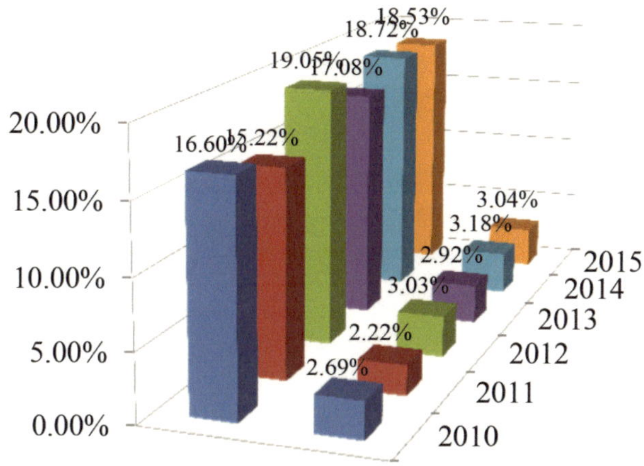

Figure 7. Contribution of cassava production quantity in Pati Regency. (Source: [14]; Secondary Data Analysis, 2017).

From these data, the contributions of 2010–2015 tend to increase, which were supported by the quality of human resources in the management of natural resources effectively and efficiently through appropriate technology utilization for determining the amount of productivity.

Therefore, it could be seen that the revitalization of tapioca industry had potential availability to meet the food demand for tapioca, which was supported by the increased production of agro-industrial areas in Pati Regency. It can be possible to be applied in other location/area considering that there still exists productive cassava farming lands at regional and national scale (**Figure 7**).

The presentation of BPS data for the last 5 years was an inventory of land resources and cassava production of certain years that were distributed all over Indonesia. It can facilitate the management of land and the use of cassava as raw material for food industry and as a crop that is easy to be cultivated. In this context, data in **Table 4** represent the existing land requirement to know the location and position for planning and direction of infrastructure development related to trade accessibility (transportation and or modes of transportation) of

Year	National		Central Java Province		Pati Regency	
	Area (ha)	Production (ton)	Area (ha)	Production (ton)	Area (ha)	Production (ton)
2010	1,183,047	23,918,118	188,080	3,876,242	21,989	643,558
2011	1,184,696	24,044,025	173,195	3,501,458	17,431	532,874
2012	1,129,688	24,177,372	176,849	3,848,462	19,696	732,961
2013	1,065,752	23,936,921	161,783	4,089,635	16,163	698,325
2014	1,149,208	23,436,384	153,201	3,977,810	17,871	744,746
2015	949,253	21,801,415	150,874	3,571,594	15,200	661,976

Source: [14, 15, 20].

Table 4. Production and cultivated areas of cassava in Indonesia.

cassava yields to industry or market. According to [21, 22], farming land has a strategic function as basic resources in land-based farming. Determination of infestation strategy based on geographical condition could illustrate the potential of a region (**Table 5**).

No	Regency/municipal	Area (ha)			Difference associative
		Normative source (BPS)		Productive	
		Regency	Province	Correction	
1	Banjarnegara	8400	8400	6403.17	−1996.83
2	Banyumas	2987	2987	1540.17	−1446.83
3	Batang	1825	1825	1791.62	−1666.62
4	Blora	2482	2482	3340.92	+858.92
5	Boyolali	5057	5057	6710.78	+1653.78
6	Brebes	1872	1872	1198.03	−673.97
7	Cilacap	4413	4381	3159.81	−1237.19
8	Demak	428	428	952.83	+524.83
9	Grobogan	1241	1272	964.38	+292.12
10	Jepara	9073	9073	8841.35	−231.65
11	Karanganyar	4324	4324	539.19	−3784.81
12	Kebumen	5436	5436	1188.45	−4247.55
13	Kendal	571	694	6121.51	5489.01
14	Klaten	704	698	6312.83	5611.83
15	Kudus	1263	1488	5801.41	4425.91
16	Magelang Regency	2070	2070	3328.18	1258.18
17	Pati	18,544	17,871	15114.17	−3093.33
18	Pekalongan	8383	504	3286.01	1157.49
19	Pemalang	1401	1415	1576.35	528.35
20	Purbalingga	3291	3304	2232.24	−1065.26
21	Purworejo	5485	5489	267.42	−5217
22	Rembang	4815	4815	775.65	4039.35
23	Salatiga	180	180	73.80	−106.2
24	Semarang Regency	1812	1822	2177.68	−342.32
25	Semarang Municipal	9318	420	3312.16	−1156.84
26	Sragen	2491	2491	483.91	−2007.09
27	Sukoharjo	1600	1600	204.83	−1395.17
28	Tegal Regency	501	517	7573.40	7064.4
29	Tegal Municipal	0	0	78.53	−78.53
30	Temanggung	1739	1739	2289.21	−51105.79

No	Regency/municipal	Area (ha)			Difference associative
		Normative source (BPS)		Productive	
		Regency	Province	Correction	
31	Wonogiri	51,656	51,656	24761.21	−26894.79
32	Wonosobo	6880	6880	382.66	−6497.34
33	Magelang Municipal	24	2	0.83	12.17
Total		170,266	153,192	122784.71	−38944.78

Source: [14, 23]; Geometric Corrections, 2017.

Table 5. Normative and productive cassava farming land in Central Java Province.

Secondary data analysis on cassava farming land area in this research, based on statistic data of BPS (Regency), showed that the normative area was 170,266 ha or in other words there was a difference in the total of 17,074 ha if compared to data source of BPS (Province) in total 153,192 ha. While the result of geometric corrections used as a sampling method showed that the productive farming land was 122784.71 ha or it was less with 38944.78 ha of normative area. The dynamics of land use had a significant effect on the production size and quantity, as well as on population and economic growth in the area.

4. Conclusions

The land existence as a sustainable resource is closely linked to the living space and surrounding of natural environments, which is influenced by the effects of weather and climate (sunlight, rainfall, wind, erosion, climate change, etc.). The intensity of land use under the forest stands could be known from the cropping pattern, which is conducted in units of a cycle time. The cropping pattern in this research was a sequence or a combination of cropping systems that were analyzed in terms of spatial and temporal dimensions on a land plot. Intensification of land use under the stands was fitted to the age of the stands, season and type of crops. The appropriate cropping pattern with biophysical conditions of Pati Regency was the intercropping on the stands that have more than 10 years old and monoculture on the stands that have less than 10 years old. Diversification of both cropping patterns requires the harmonization efforts between the shade effect and characteristics of the agricultural crops. Productivity of the crops was equivalent to 75% of land without shade and 50% of land with shade.

Acknowledgements

This chapter was written based on result of National Strategic and research grant on scheme of Masterplan Economic Development Acceleration and Expansion of Indonesia (MP3EI), which was funded by the Directorate of Research and Community Service (DRPM), Ministry of Research, Technology and Higher Education of Indonesia.

Thanks to: Director of Research and Community Service, General Director of Research and Development, Rector of Universitas Negeri Semarang (UNNES), Chairman of Research and Community Service Institute of UNNES, Dean of Faculty of Social Sciences of UNNES, Head of Forest Resources Bureau (Perhutani) Unit I of Central Java and all farmers (*pesanggem*) of Muria Forest who have supported and worked together to support this research.

Conflict of interest

I declare that I have no conflict of interest within this research or this manuscript.

A. Appendix 1

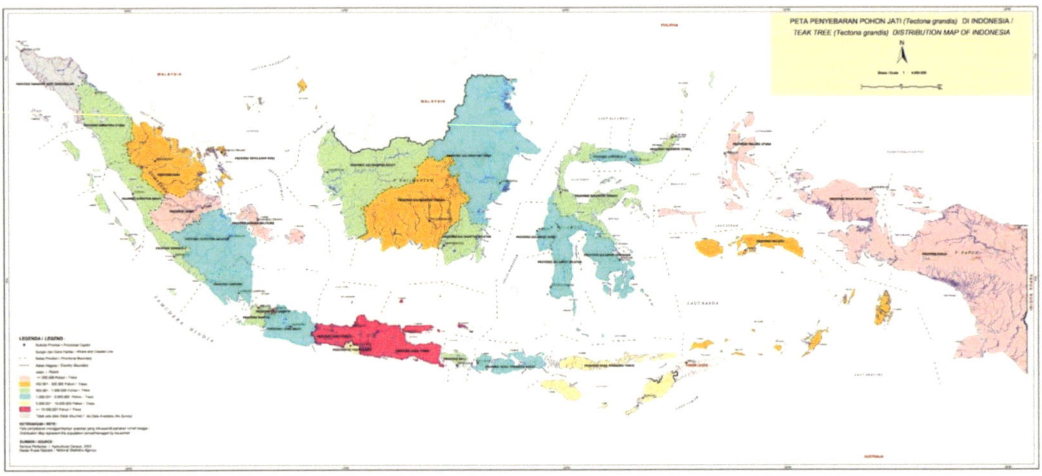

Figure. Teak tree distribution map of Indonesia (Source: webgis.dephut.go.id).

Author details

Eva Banowati* and Satya Budi Nugraha

*Address all correspondence to: evabanowatigeografi@mail.unnes.ac.id

Universitas Negeri Semarang, Semarang, Indonesia

References

[1] Johnson C. An Introduction to Agroforestry [Internet]. 2006. pp. 279-281. Available from: http://linkinghub.elsevier.com/retrieve/pii/037811279590008X

[2] Jose S. Agroforestry for ecosystem services and environmental benefits: An overview. Agroforestry Systems. 2009;**76**(1):1-10

[3] Atangana A, Khasa D, Chang S, Degrande A. Agroforestry and biodiversity conservation in tropical landscapes. In: Tropical Agroforestry. Netherlands: Springer Science+Business Media B.V. 2014. pp. 227-232

[4] Banowati E. Pemberdayaan penduduk pesanggem untuk pengentasan kemiskinan dan percepatan pemulihan sumberdaya hutan Muria. Jurnal SPATIAL: Wahana Komunikasi dan Informasi Geografi. 2016;**16**(2):39-46

[5] Hairiah K, Sardjono MA, Sabarnurdin S. Pengantar Agroforestri. Bogor: World Agroforestry Center (ICRAF); 2003. 44p

[6] Banowati E. Pembangunan Sumber Daya Hutan Berbasis Masyaraakt di Kawasan Hutan Muria-Kabupaten Pati. Yogyakarta: Faculty of Geography UGM. 2011

[7] Nursanti I. Tanggap tanaman jagung terhadap pemberian pupuk produk samping cair pabrik kelapa sawit plus zeolit. Jurnal Media Pertanian. 2016;**1**(2):62-68

[8] Suharjito D. Penerapan Multisistem Silvikultur Pada Areal Hutan Produksi: Tinjauan Sosial Budaya Masyarakat Lokal. Proceeding of National Workshop. 2010

[9] Pattinama MJ. Pengentasan kemiskinan dengan kearifan lokal (Studi kasus di pulau Buru-Maluku dan Surade-Jawa Barat). Jurnal Makara Sosial Humaniora. 2009;**13**(1):1-12

[10] Mustofa MS. Perilaku masyarakat desa hutan dalam memanfaatkan lahan di bawah tegakan. KOMUNITAS: International Journal of Indonesian Society and Culture. 2011; **3**(1):1-11

[11] Banowati E, Prajanti SDW. Developing the understand cropping system (PLDT) for sustainable livelihood. Management of Environmental Quality An International Journal [Internet]. 2017;**28**(5):769-782. DOI: 10.1108/MEQ-08-2015-0163

[12] Murniyanto E, Sugito Y, Jati T. Potensi *X. sagittifolium* di bawah tegakan hutan produksi jati: Penunjang ketahanan pangan. In: Reformasi Pertanian Terintegrasi menuju Kedaulatan pangan. Proceeding of National Conference. 2011. pp. 1-8

[13] Ahmed J, Khan HW. Biodiversity Conservation on Rural Development in the Aga Khan Rural Support Programme [Internet]. 1993. Available from: http://www.akdn.org/our-agencies/aga-khan-foundation

[14] Badan Pusat Statistik Kabupaten Pati. Pati Regency in Figures 2016. Pati, Central Java, Indonesia: Badan Pusat Statistik Kabupaten Pati; 2016. 316p

[15] Badan Pusat Statistik Kabupaten Pati. Indicators of Indikator Pertanian Kabupaten Pati. Pati, Central Java, Indonesia: Badan Pusat Statistik Kabupaten Pati; 2015. 152p

[16] Muhadi. Kajian pengembangan strategi potensial industri tepung tapioka rakyat (ITTARA) di Kabupaten Lampung Timur. Jurnal Teknologi Industri dan Hasil Pertanian 2017; **22**(1):52-62

[17] Fuglie KO. Investing in agricultural productivity in Indonesia. Forum Penelit Agro Ekon. Journal of Forum Penelitian Agro Ekonom. 1999;**17**(2):1-16

[18] Warr P. Agricultural productivity, poverty and inequality in Indonesia. In: 60th Annual Conference on Australian Agricultural & Resource Economics Society; 2016. pp. 1-6

[19] Hendriadi A. Agricultural productivity in Indonesian provinces. In: Giap TK, Merdikawati N, Amri M, Kong YT, editors. 2015 Agricultural Productivity, Decentralisation and Competitiveness Analysis for Provinces and Regions of Indonesia. Singapore: World Scientific; 2016. p. 126

[20] Badan Pusat Statistik Kabupaten Pati. Pati in Figures. Pati, Central Java, Indonesia: Badan Pusat Statistik Kabupaten Pati; 2014. 316p

[21] Gunawan J, Fraser K. Exploring young and green entrepreneurship in Indonesia: An introduction. Journal of Asian Business Strategy. 2016;**6**(9):185

[22] Gunawan J, Fraser K. Green jobs in Indonesia: Potentials and prospects for national strategy. In: Economy of Tomorrow. Jakarta: Friedrich Ebert Stiftung. 2014. p. 31

[23] Badan Pusat Statistik Provinsi Jawa Tengah. Jawa Tengah Province in Figures 2016. Semarang, Central Java, Indonesia: Badan Pusat Statistik Provinsi Jawa Tengah; 2016. 411p